Lunar and Planetary Cartography in Russia

More information about this series at http://www.springer.com/series/5664

Vladislav Shevchenko • Zhanna Rodionova •
Gregory Michael

Lunar and Planetary Cartography in Russia

Springer

Vladislav Shevchenko
Lunar and Planetary Research
Sternberg Astronomical Institute
Lomonos
Moscow, Russia

Zhanna Rodionova
Lunar and Planetary Research
Sternberg Astronomical Institute
Lomonos
Moscow, Russia

Gregory Michael
Freie Universität Berlin
Berlin, Germany

ISSN 0067-0057 ISSN 2214-7985 (electronic)
Astrophysics and Space Science Library
ISBN 978-3-319-21038-4 ISBN 978-3-319-21039-1 (eBook)
DOI 10.1007/978-3-319-21039-1

Library of Congress Control Number: 2015950753

Springer Cham Heidelberg New York Dordrecht London

Cover illustration: Venus Relief Map (Lazarev et al. 2012). See also Fig. 7.10.

Printed on acid-free paper

Springer International Publishing AG Switzerland is part of Springer Science+Business Media (www.springer.com)

Preface

The stages of development of the theory and practice of lunar and planetary cartography in Russia have been directly related to the progress in the field of space research beginning in the second half of the twentieth century. Various overview maps of the lunar nearside were compiled in the pre-space era, generally containing information on the geological structure of the lunar surface, primarily as preparatory material for site selection for detailed research using spacecraft.

Work on the compilation of lunar maps in Russia began in earnest after the successful execution of the epochal space experiment to photograph the farside of the Moon. We may recall that the work on the first maps to include the lunar farside as well as the first globes was considered not only a highly important scientific achievement but also a project of national significance proceeding on the basis of governmental decisions allocating significant funding.

The greatest specialists of the time participated in the technical implementation of the project to photograph the lunar farside. Chief Designer Sergei Pavlovich Korolev led the development of the rocket launcher systems. The complex trajectories of lunar flyby with later returns to Earth were elaborated in a team headed by the leading scientist of the field, Boris Viktorovich Rauschenbach.

On the recommendation of the world-famous astrophysicist, Iosif Shklovsky, the cartographic processing and subsequent image analysis was assigned to a leading scientist of the Sternberg Astronomical Institute (GAISh), Yury Naumovich Lipsky. It is worth noting that Shklovsky's group were among the first astronomers to be included in the development of experiments in space. They developed and realised the "artificial comet" experiment, which gave rise to the technique of determining a spacecraft trajectory by observing its position close to the Moon. Before this, *Luna 1*, which was supposed to have impacted the western region of the lunar surface, passed by the Moon at over 6000 km from the surface, lacking the ability to make a trajectory correction. Having the "artificial comet" experiment on board *Luna 2* ensured an accurate impact into the Moon. The radio tracking techniques used today had not, at that time, reached the necessary levels of precision.

The compilation of the first map of the lunar farside required the solution of new technical problems: the construction, for example, of a new cartographic projection and the determination of selenographic coordinates of many features within a territory which was previously unknown. A partial solution to the problem was inherent in the parameters of the experiment (see Chap. 1). The flight path of *Luna 3* was designed so that not the whole of the far hemisphere was visible from the point of observation, leaving a part of the image covering the eastern edge zone of the Moon. This allowed the coordinate system of known features to be extended into the previously unseen area. At the same time, the observation point was chosen to be able to cover as much as possible of the farside.

This last requirement included an insoluble contradiction. Astronomers knew that the unique reflective properties of the lunar surface were such that the maximum of the reflected light is returned in the direction of the source. In this case, the so-called full-Moon effect occurs, when shadows cast by any relief disappear, and the resulting image becomes "blind": it is possible only to see variations in surface albedo—the reflective characteristic of surface features. The experiment designers' desire to encompass as much territory as possible in their first photographs of the farside resulted in the majority of the image having full-Moon conditions. Lipsky and his group encountered difficulties in the identification of known structures. Despite the use of specially developed enhancement procedures, enabling them to find many dozens of craters, not all of these were re-identified in later images with different illumination conditions.

Chapter 1 shows several variants of the processing of the first images from the farside of the Moon, with detailed analysis of each one.

Nevertheless, the first experience of space photography of the Earth's satellite was recognised as a success, and a government decision was taken to carry out a global survey of the lunar surface with a new space experiment. Six years later, in 1965, *Zond 3* made a new flight around the Moon, sending back images of practically the whole of the lunar globe.

In Chap. 2, we examine the preparation of several editions of a *Complete Map of the Moon* and complete lunar globes: the results of the global survey. At this point of development of Soviet lunar cartography, the astronomers were joined by professional cartographers, in particular from the USSR Topogeodesic Service, bringing their decades of experience of working with cartographic images of the natural surface of the Earth.

In Chap. 3, covering the cartography of the lunar nearside, it is worth noting in particular the description of the large-scale maps of the equatorial zone. This edition, aside from its scientific value, was widely utilised for engineering applications, particularly as an information source for the selection of landing sites for both unmanned and manned spacecraft.

With respect to Chap. 4, it may be noted that in Earth cartography, the fraction of thematic and complex atlas material makes up about 80 % of published material. The experience of compilation of maps for the Earth shows that the cartographic modelling of the diverse natural properties of the environment is an essential element of studying and understanding them. This experience is fully applicable to the study of the nature of planets and satellites in the Solar System. At present, thematic lunar cartography is the most developed among these. Selenographic, gravimetric, hypsometric, structural and morphological, spectral, polarimetric and other types of thematic cartography reflect the characteristics of the nature and structure of the lunar surface. At the same time, the cartographic method of research permits the study of processes of evolution of particular structures and reveals the spatial relationships between the structures and the phenomena which produce them.

As for other fields of cartography, lunar thematic cartography achieved its greatest development simultaneously with the development of the corresponding space technologies. The source of information for the various maps is primarily remote sensing by spacecraft instruments. Spacecraft observations can utilise the entire range of natural and reflected radiation of the Moon. Gamma ray emission, for example, of lunar surface rocks provides data on the presence of natural radioactive elements: potassium, thorium and uranium. Measurement of the x-ray fluorescence of the lunar soil allows the determination of the content of magnesium, aluminium and silicon. Images of solar light in the visible range reflected from the surface are a universal method. These data are used to construct photometric, brightness, polarisation and zoned spectral maps of the surface, which themselves give information about features of the chemical composition and mechanical properties of the surface layer of lunar regolith. Today it is evident that thematic cartography of the Moon and planets needs to be carried out by a large group of specialists in a coordinated programme, to result in a series of maps and atlases offering a complete characterisation of an integrated group of phenomena.

An essential part of the information content of a map consists of the individual names of surface features, which enable both the search for and reference to particular structures. Each name is a short code for a feature's location, characterising its structure, while also attaching a cultural or historical meaning. The toponomy of the Moon has the longest history among extraterrestrial bodies and is employed in Chap. 5 as an example to trace the development of a planetary toponomy and the typical problems which are encountered in naming planetary surface features.

In recent times, not only the Moon but many other bodies of the Solar System have been intensively studied using spacecraft. In Russian lunar and planetary cartography, it has been Mars, with its satellites, and Venus where the most development has occurred in the production of maps, globes and atlases. Besides general geographical maps for these bodies, geomorphological and tectonic maps were also produced. The source materials for the products were the data returned by spacecraft. In the last 30 years, five interplanetary spacecraft were sent to Venus by different national space agencies, while, at the same time, 20 spacecraft of various countries visited Mars and its satellites. The preparation of cartographic products needed to bring together specialists from various fields. As well as the astronomers from the Sternberg Astronomical Institute involved in the thematic mapping of planets and moons, there were also cartographers, geologists and geochemists from various academy, university and specialised Russian organisations. These cartographic materials and various particulars of their production are described in Chaps. 6 and 7.

Moscow Vladislav Shevchenko
Moscow Zhanna Rodionova
Berlin Gregory Michael

Abbreviations

The following is a list of abbreviations used in the book. Many Russian institutions are commonly known by pronounced abbreviated forms, and some have well-known alternative abbreviations for English translations of their names. To avoid confusion, this book uses transliterated forms of the Russian abbreviations throughout.

COSPAR	Committee on Space Research
ESA	European Space Agency
GAISh	Sternberg Astronomical Institute (SAI) *(Gosudarstvennyy Astronomicheskiy Institut imeni P.K. Shternberga)*
GEOKhI	Vernadsky Institute of Geochemistry and Analytical Chemistry *(Institut geokhimii i analiticheskoy khimii imeni V.I. Vernadskogo)*
GUGK	General Directorate of Geodesy and Cartography *(Glavnoye upravleniye geodezii i kartografii)*
IAU	International Astronomical Union
IKI	Space Research Institute *(Institut Kosmicheskikh Issledovaniy)*
MGU	Lomonosov Moscow State University (MSU) *(Moskovskiy gosudarstvennyy universitet imeni M. V. Lomonosova)*
MIIGAiK	Moscow Institute of Geodesy and Cartography *(Moskovskiy gosudarstvennyy universitet geodezii i kartografii)*
NASA	National Aeronautics and Space Administration (USA)
NPO Energomash	Energomash Scientific and Production Association *(Nauchno-proizvodstvennoye ob"yedineniye Energomash)*
PKO Kartografiya	Cartography Production Mapping Association *(Proizvodstvennoye kartosostavitel'skoye ob"yedineniye «Kartografiya»)*
TsNIIGAiK	Central Scientific Research Institute for Geodesy, Aerial Photography and Cartography *(Tsentral'nyy nauchno-issledovatel'skiy institut geodezii, aeros"yomki i kartografii)*
WGPSN	Working Group for Planetary System Nomenclature (IAU)

Acknowledgement

The authors express their sincere thanks to Renato Dicati for the suggestion to write the book, to the members of the Department of Lunar and Planetary Research at GAISh, to the students and staff of the Department of Cartography and Geoinformatics at the MGU Geography Faculty, to the students and staff of the Laboratory of Planetary Cartography at MIIGAiK and to all those who have participated in the construction of spacecraft missions and the processing of the returned data, both in Russia and in other countries.

Contents

Chapter 1
The First Maps, Atlas and Globus of the Farside

The Moon should, in the near future, become a part of the world civilisation. Its population will, of course, be few. Yet stable outposts will appear for the advancement of science.
B. Chertok, 2011.

Abstract The first *Map of the Lunar Farside* at 1:10,000,000 scale was published in the Soviet Union on the basis of images of the *Luna 3* spacecraft launched from the Baikonur cosmodrome on 4th October 1959. New methods for processing the original photographic images developed by Yu. N. Lipsky largely eliminated the radio interference enabling the identification of many real features of the lunar terrain. The newly discovered surface features were compiled onto a map using a unified selenographic coordinate system. The *Atlas of the Lunar Farside*, published in 1960, includes 30 photographs taken by *Luna 3*, the results of their analysis and a catalogue with detailed descriptions of 729 identified features. The first globe, showing the nearside and a part of the farside. was produced in 1961.

Academician Boris Evseyevich Chertok – the legendary spacecraft engineer, comrade-in-arms and deputy of Chief Designer Sergei Pavlovich Korolev – described in his memoirs of the distant events of October 1959 the moment of receiving the first images from the farside of the Moon (Chertok 1996, 2006):

I took a place beside Boguslavsky at the apparatus for recording on electrochemical paper.

From the receiving station they reported:

— Range: fifty thousand. Signal stable. Receiving!

The command was given to produce the image. Again the responsibility was with FTU.

Line by line on the paper, a grey image began to appear. A disc which, with sufficient imagination, details could be distinguished.

Impatiently, Korolev burst into our cramped little room.

— Well, what have you got there?

— We've found that the Moon is round — I said.

Boguslavsky pulled the recorded image from the apparatus, showed it to Korolev, and calmly tore it up. Sergei Pavlovich didn't raise an eyebrow.

— Why so soon, Evgenii Yakovlevich? It's just the first, you know – just the first!

— It's bad – such a muddy image. Now we'll get rid of the interference, and the next frames will be OK.

V. Shevchenko et al., *Lunar and Planetary Cartography in Russia*, Astrophysics and Space Science Library 425, DOI 10.1007/978-3-319-21039-1_1

One after the other, the images appeared on the paper, each more distinct than the last. We cheered and congratulated each other. Boguslavsky reassured us that on the photographic film, which was being processed in Moscow, everything would be much better.

1959 became the beginning of a new era in lunar mapping. As far as we know, there were no maps of the Moon published in Russia before 1960. However, the world's first *Map of the lunar farside* was published in the USSR. The Soviet interplanetary spacecraft *Luna 3* imaged the Moon for 40 minutes on 7th October 1959 (Pervye fotografii obratnoy storonu Luny 1959) (Figs. 1.1 and 1.2). As a part of the development of the technology for the photography and image transmission, it was necessary to build an orientation system, consisting of optical and gyroscopic sensors, electronic logic units, and steering thrusters to point the spacecraft in the required direction. The photography was done using a camera with two objectives, having different focal lengths. The film was developed, fixed, rinsed and dried automatically on board. The image was transmitted on command from Earth to a ground-based receiving station (Mikhailov and Straut 1968). Images of the eastern part of the nearside and the western part of the farside were transmitted by radio to Earth (Fig. 1.3).

Decoding the received images of the lunar surface, as well as the construction of the first lunar farside map, was simultaneously and independently carried out by three organisations: in Moscow, at Pulkovo, and in Kharkov. In Moscow, this work

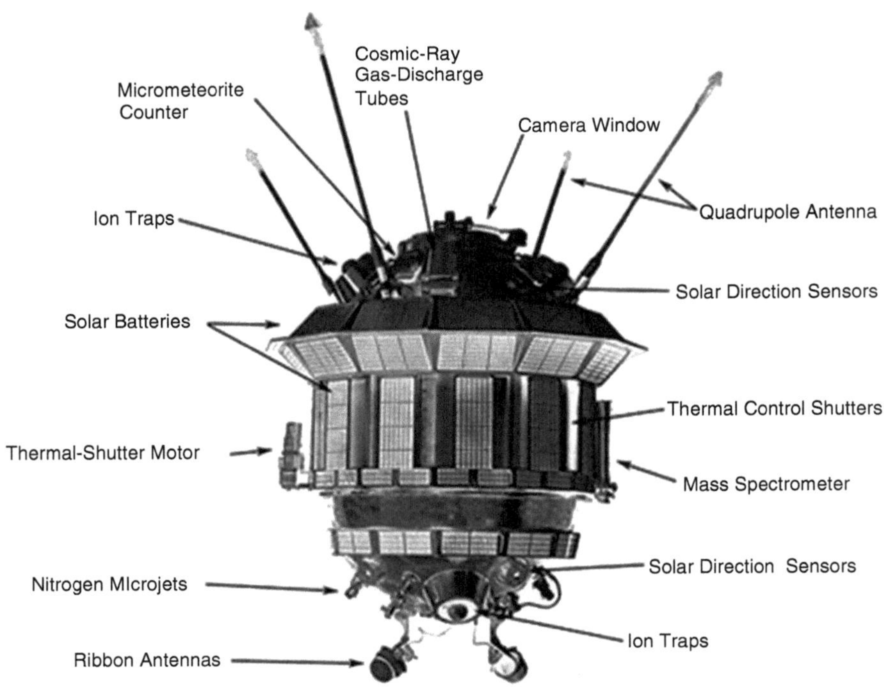

Fig. 1.1 Interplanetary spacecraft *Luna 3*

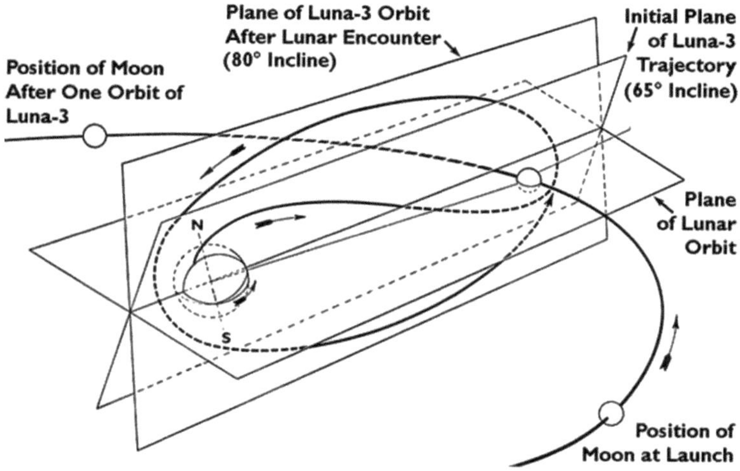

Fig. 1.2 Flight trajectory of *Luna 3*

was lead by Yu. N. Lipsky (Sternberg Astronomical Institute) and N. A Sokolova (Central Science-research Institute of Geodesy, Aerial Photography and Cartography); in Pulkovo, by A. V. Markov (Main Astronomical Observatory of the USSR Academy of Sciences) and in Kharkov, by N. P. Barabashova (Astronomical Observatory of the A. M. Gorkii Kharkov State University).

Figure 1.4 shows the *Map of the lunar farside* at 1:10,000,000 scale (GAISh MGU and TsNIIGAiK). Details of the lunar farside seen for the first time were indicated with an arbitrary notation: their coordinates were determined in a unified selenographic coordinate system, and 18 large features were given names approved by the IAU. Figure 1.5 shows the map produced at Pulkovo Observatory. The map of Fig. 1.4 shows significantly more details because of the use of a special method of image decoding and processing developed at the Sternberg Astronomical Institute (GAISh MGU). Because there was practically no computing technology available to process images, Yu. N. Lipsky developed a complex technique for photographic transformation of the original images. The method enabled much of the radio noise to be removed, revealing many more real features of the surface. The hand drawn map produced on the basis of this work is shown in Fig. 1.6. However, Kopal and Carder preferred the map from Pulkovo (Fig. 1.5):

"In some respects, the schematic map compiled at Pulkovo is superior to the Sternberg map in that an attempt was made to reproduce both the shapes and brightness of the areas of different reflectivity" (Kopal and Carder 1974).

The Pulkovo map (Breido and Shchegolev 1963), as well as showing areas of different reflectivity, includes 107 of the most distinctly visible features, of which 56 are found on the farside. The map included photometric estimates of the reflectivity of features, which were made by A. V. Markov using Luna 3 images of the full lunar disc. This was the first experience of photometric research from space

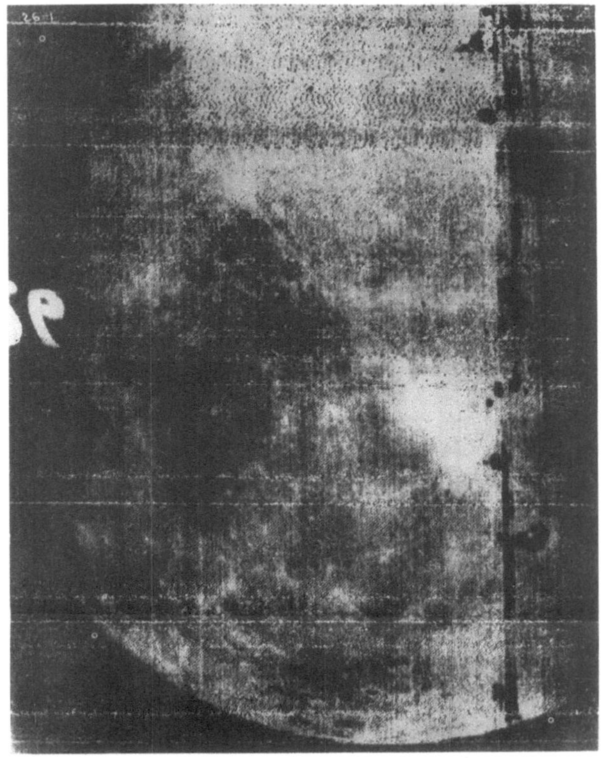

Fig. 1.3 Photograph of the Moon obtained by *Luna 3*

observations. His work made first use of the "reference catalogue method" for photometric calibration of images, which became widely used in later works (Markov 1963).

Figure 1.7 shows the map produced at Arizona University, also based on images from *Luna 3* (Whitaker 1963). The author used an original technique for reprojecting the photographs onto a spherical screen developed in the university's Lunar and Planetary Laboratory. The map is notable for the high level of detail in the representation of individual features, especially close to the terminator and in parts of the nearside visible from the spacecraft close to the 150°E meridian. The light aureole around some craters and bright ray systems are shown particularly carefully. Whitaker's map is intermediate in the level of detail between that of GAISh MGU and TsNIIGAiK and the generalised scheme of reflectivity from Pulkovo (Shevchenko 1986).

A detailed description of the apparatus, methods of receiving the images, decoding and processing is given in the Atlas of the Lunar Farside (1960). The diameters of the original large and small-scale images of the lunar disc received by the photoregistrators on Earth were 25 and 10 mm, respectively. Positives on photographic paper and film were also used, produced by apparatus decoding the spacecraft signal, which have been recorded in magnetic tape. On these positives

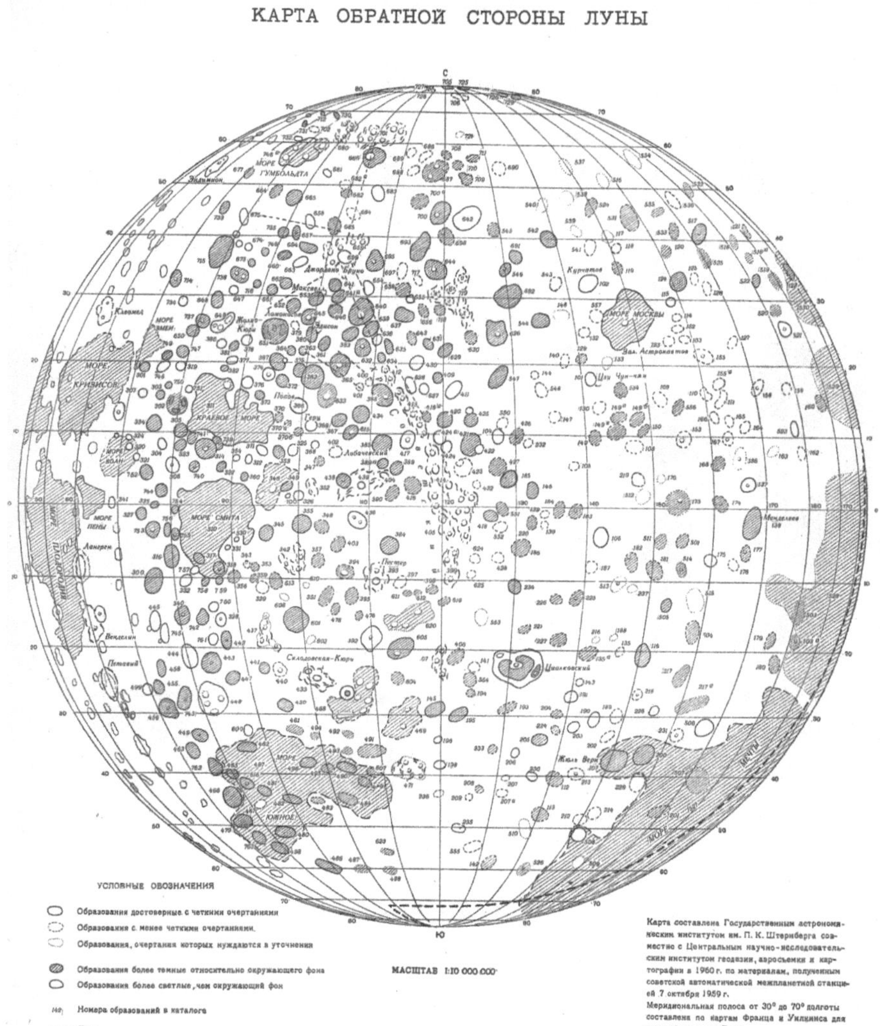

Fig. 1.4 Map of the lunar farside compiled in Moscow at GAISh MGU and TsNIIGAiK (1960)

the diameter of the lunar disc was 100 mm for the small-scale photographs and about 250 mm for the large scale. Multiple recording of the images in special operating modes of the apparatus (photometric slices) permitted an eventual increase in the fidelity of decoding by the Moscow group. The photometric slices method involved multiple amplifications of the contrast of regions on the negative with similar photometric properties. The radio signals received from *Luna 3* were severely distorted by interference. Furthermore, the axes of the camera objectives

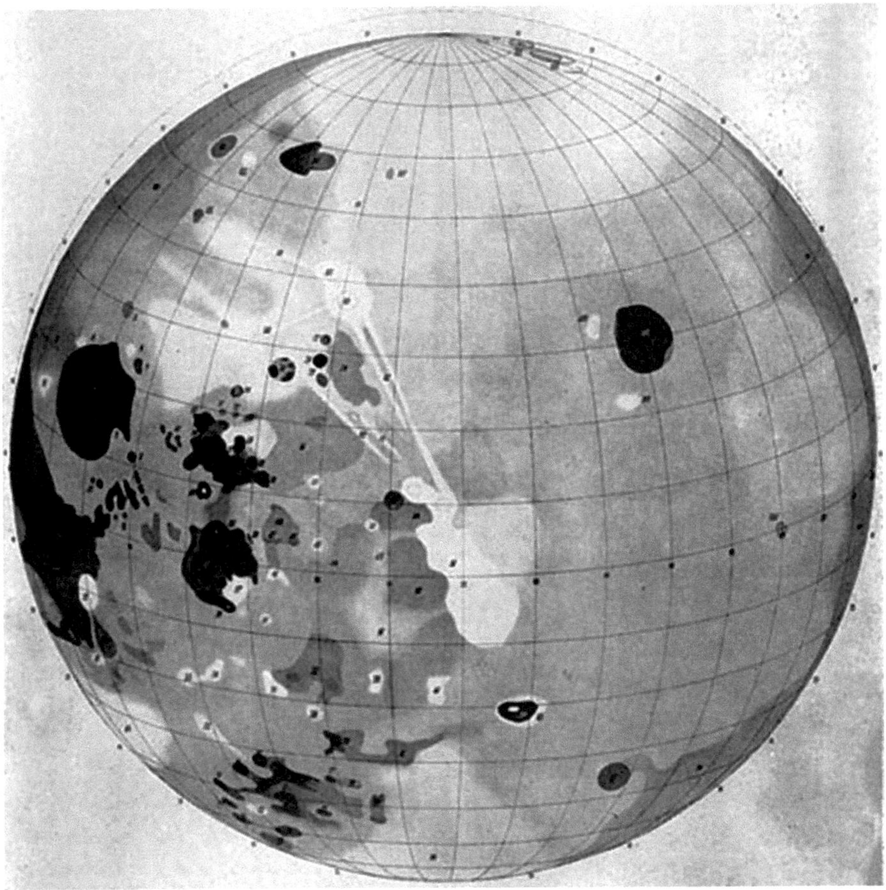

Fig. 1.5 Map of the lunar farside compiled at Pulkovo observatory (1963)

were nearly aligned with the direction of the solar illumination of the lunar surface, and thus the contrast variation of surface relief features was minimal.

An image, processed with modern methods, is shown in Fig. 1.8.

In the process of compilation of the map (Fig. 1.4), the surface features were divided into three categories. The first included features with distinct outlines, which were well identifiable on three or more frames, as well as all features of the edge zone of the nearside hemisphere. The second category included features which were identifiable only into two frames, and the third category included features with indistinct outlines. The map indicated structures which were darker or lighter than the surrounding regions, as well as ray systems, with special marks. A coordinate grid in external perspective projection was constructed using the known position of the spacecraft and its distance from the Moon, and this was used to connect features of the visible side with the newly identified features of the farside. In planning the experiment, it was proposed that parts of the East nearside

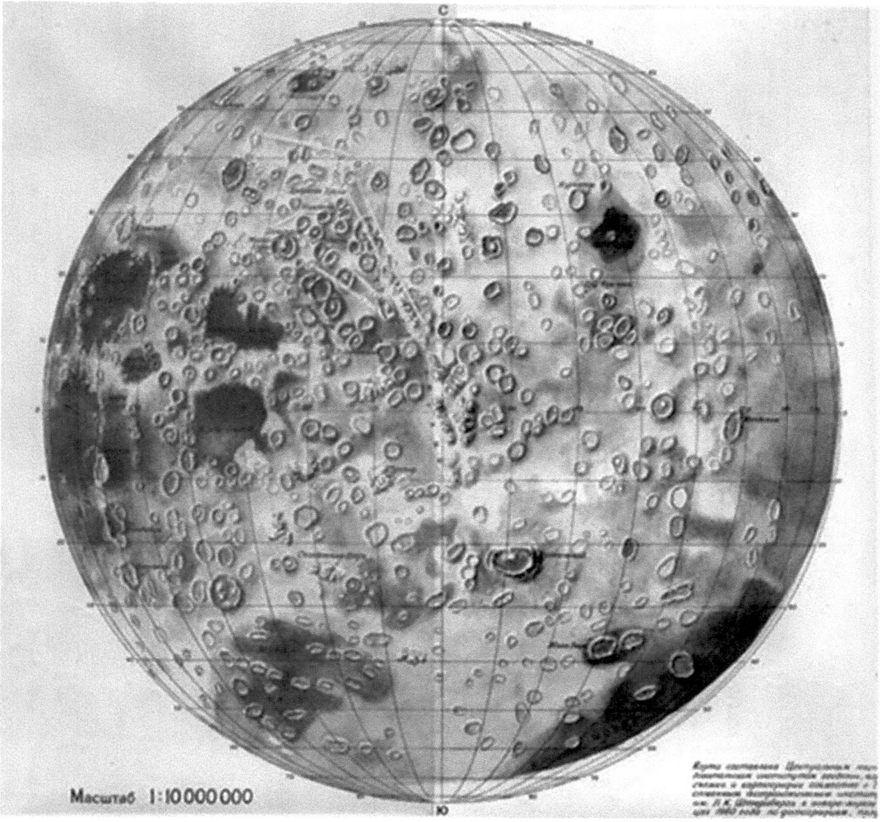

Fig. 1.6 Half-tone variant of the lunar farside map compiled in Moscow at GAISh MGU and TsNIIGAiK (1963). Details of relief were drawn by artist-cartographers in the typical manner for lunar maps

and the libration zone should be included so that known features of the lunar surface could facilitate a connection of the nearside coordinate system with the unknown features of the farside.

The final map of the lunar farside was constructed in equatorial orthographic projection with a central meridian of 120°E. The coordinate grid was drawn with a 10°interval. The diameter of the lunar hemisphere on the map was 34.76 cm. The map, as included in the Atlas of the Lunar Farside, was divided into four parts, which had some overlap. The source material for the eastern part of the nearside (the west part of the map) was the map of Wilkins and Moore (1955). This zone includes Mare Humboldtianum, Craters Endymion and Cleomedes, Mare Anguis, Mare Crisium, Mare Marginis, Mare Smythii, Mare Undarum, Mare Spumans, Mare Fecunditatis, craters Langrenus, Vendelinus, Petavius, Mare Australe. On the farside are shown the craters: Giordano Bruno, Maxwell, Lomonosov, Edison, Joliot, Jules Verne, Hertz, Popov, Lobachevskii, Pasteur, Tsu Chung-Chi,

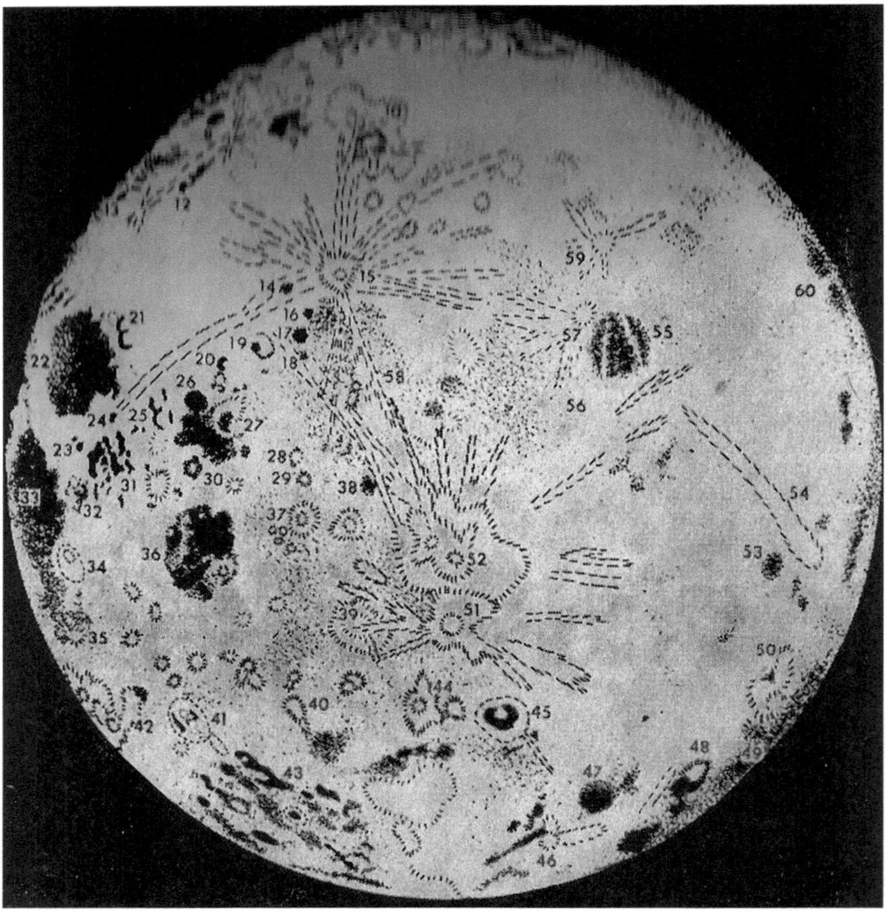

Fig. 1.7 Map of the lunar farside compiled from *Luna 3* materials at Arizona University (1963)

Mendeleyev, Tsiolkovskii, Sklodowska-Curie, Kurchatov as well as Mare Moscoviensis, Mare Ingenii, Sinus Astronauticus. A light extended feature which, by analogy with several details of relief on the nearside, was identified as a mountain structure was given the name *Montes Sovietici*. However, later images with different illumination conditions did not confirm the presence of a mountain ridge, and the name was removed from the nomenclature. Likewise, the name Sinus Astronauticus was also not retained, although corresponding morphological structures were found in later images.

On the basis of results from the first imaging of the lunar farside, the morphological characteristics of many remarkable structures of the lunar surface were identified.

The *Luna 3* images revealed the extensive ray system of Giordano Bruno Crater. Special photometric slices revealed the complex structure of the floor of

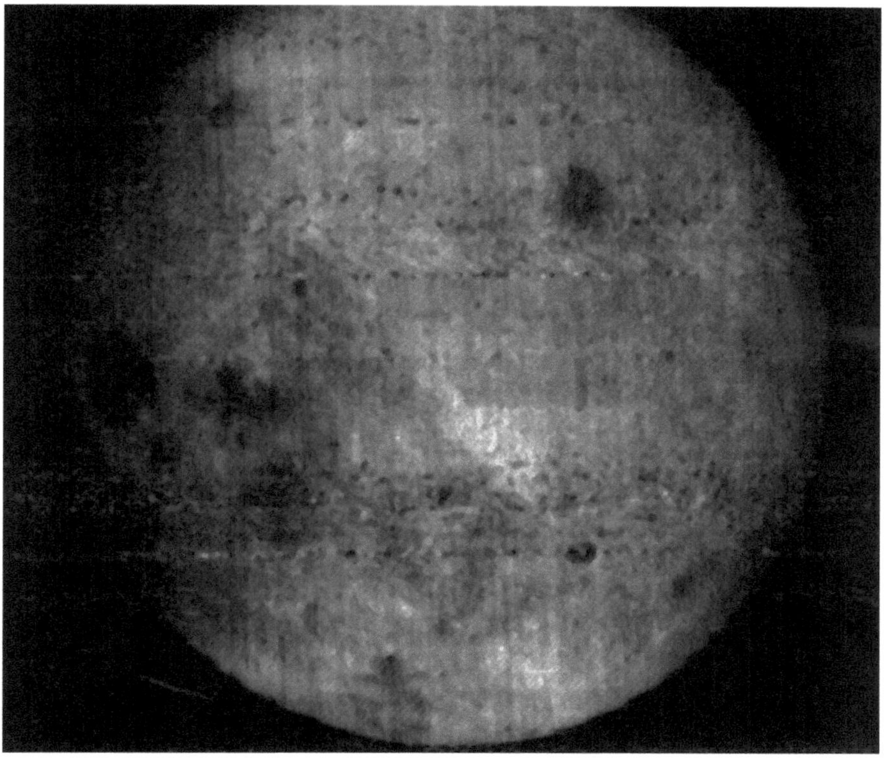

Fig. 1.8 Image of the lunar farside from *Luna 3* processed with modern methods

Tsiolkovskii Crater, including its central peak. It is interesting to know that the contour of the western boundary of the structure named Mare Ingenii on this map corresponds with the western edge of South Pole—Aitken basin on modern maps. The map also highlighted formations which are darker relative to the surrounding crater wall. In other cases, newly discovered light ray systems are shown. Since, as mentioned above, the first map of the lunar farside was constructed in orthographic projection with a central meridian chosen to be 120°E, Mare Marginis, Mare Smythii and Mare Humboldtianum, which are in the libration zone, are shown with minimal distortion.

On the initiative of the IAU, an international symposium, *Moon,* was held in 1960 in Leningrad (Pulkovo), attended by the most prominent specialists of planetology. The participants were shown the newly printed first *Atlas of the Lunar Farside* edited by N. P. Barabashov, A. A. Mikhailov and Yu. N. Lipsky: (Barabashov et al. 1960).

The Nobel Laureate Harold Urey (USA) remarked "The photographs of the far side of the Moon are, first of all, outstanding research of interest not only to scientists. It is gratifying that man has successfully accomplished this goal and revealed what was hitherto an unseen part of the Moon. From a purely scientific point of view, these images significantly change our understanding of the Moon. An interesting fact is the absence of large maria. It is possible that this shows that the maria are the result of collisions of large bodies with the lunar surface..."

Professor of Manchester University (UK), Zdeněk Kopal, said "The photographs of the lunar farside obtained by the Soviet spacecraft in October 1959 are without doubt the greatest contribution to the astronomy of the Solar System in the last decade, if not of our whole generation. They are an outstanding achievement, for which the Soviet scientists should be congratulated..."

Karol Kozel, professor of Krakow University (Poland), said "As chairman of the International Astronomical Union commission for research of the motion and figure of the Moon, I give my heartfelt congratulations to those Soviet scientists, who realised the long held dream of humanity to study the farside of the Moon. This wonderful achievement of Soviet science deserves high praise throughout the world. New possibilities for fascinating research have been opened up to astronomers, which previously revolved around untested hypotheses..." (Linder 1961).

In the *Atlas of the Lunar Farside* are 30 photographs which were made by *Luna 3*. There are also photometric slices resulting from additional processing of the images, descriptions of the source materials and processing methods. The *Map of the Lunar Farside* was first published in the atlas at a scale of 1:10,000,000, and a catalogue with detailed descriptions of the 729 newly identified relief features of the lunar farside (Fig. 1.9).

The Lunar Globe at scale 1:13,600,000, compiled under Yu. N. Lipsky at GAISh and TsNIIGiK, was hugely popular, both in the USSR and abroad. The globe

Fig. 1.9 USA science attache Schweizer (*centre*) presents pictures of the Ranger spacecraft to Yu. N. Lipsky (*left*) and P. G. Kulikovsky (*right*)

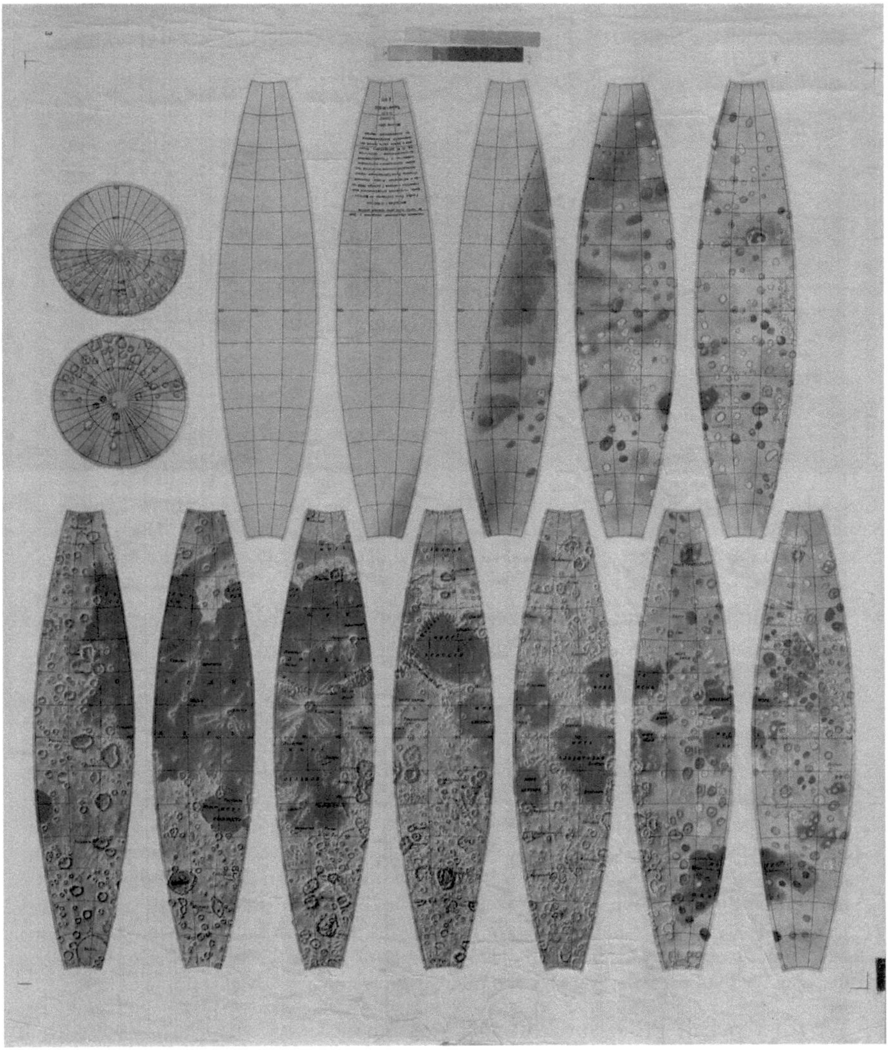

Fig. 1.10 Gores of the lunar globe showing the nearside and a part of the farside were produced in 1961 at GAISh and TsNIIGiK

showed the nearside of the Moon as well as the part of the farside which had been photographed by *Luna 3* (Fig. 1.10). As a consequence, it was reproduced in the UK at 1:10,000,000 scale.

The best images obtained by the *Lunar Reconnaissance Orbiter* (*LRO*), launched in 2009, were used to recreate the view of the Moon seen by *Luna 3* in 1959. Figure 1.11 shows both views together. The general features are identical.

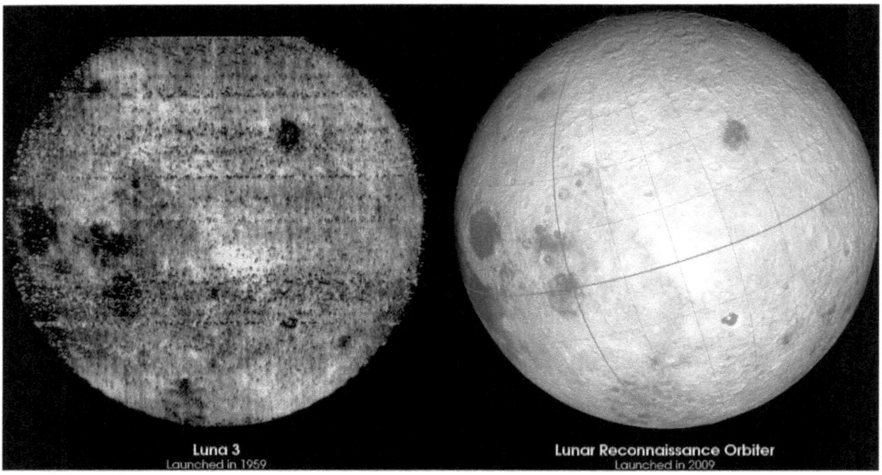

Fig. 1.11 Comparison of the first ever photograph of the lunar farside from *Luna 3* and a visualisation of the same view using *LRO* data (http://svs.gsfc.nasa.gov/goto?4109). The thicker *blue lines* on the *LRO* image indicate the equator and the 90° meridian

To conclude this chapter, we recall an interesting episode relating to receiving the *Luna 3* photographs described in Boris Chertok's book:

> Humankind admired us, and the entire Soviet population was proud of us without knowing our names. But we didn't grumble over that. "It wasn't just humankind that appreciated our achievement", said Sergey Pavlovich [Korolev], "but also a wealthy French winemaker. He announced that he would give a thousand bottles of champagne to the ones who reveal the farside of the Moon. He was certain that we wouldn't come up with anything and wasn't afraid of the risk. But once he lost, he kept his word. Of course, there was a hitch. The vintner asked the embassy in Paris to let him know where to send the champagne. The embassy was at a loss and asked our Ministry of Foreign Affairs. After multilevel coordination, the ministry gave instructions to send the bottles to the Academy of Sciences presidium. So, now, we have the honor of receiving several dozen bottles of champagne from the Academy's stock. You'll snag a couple of bottles each, and the rest will be dispersed among the Party bigwigs and others who weren't involved".

We approached Professor Philippe Masson of Université Paris-Sud with the request to find out the name of the vintner. This was his reply:

> I am pleased to inform you that in answer to my request, the Association of Champagne winemakers (Union des Maisons de Champagne) provided me the following information today:
>
> *"In 1957, French winemaker Henri Maire bet the Soviet consul a thousand bottles of champagne that Soviet satellites would not be able to see the dark side of the Moon. When photos of the Moon, taken by the Luna-3, appeared in all newspapers around the world, the winemaker acknowledged his defeat and sent a thousand bottles to the Academy of Sciences of the USSR".*

References

Barabashov, N. P., Mikhailov, A. A., & Lipskiy, Yu. N. (1960). *Atlas obratnoy storony Luny. 1960. Pod redaktsiey* (149 s). Moskva: Izdatelstvo Akademii nauk SSSR.

Breido, I. I., & Shchegolev, D. E. (1963). *Karta-shema obratnoy storony Luny. V kn. Novoe o Lune* (s. 19–27). Izdatelstvo AN SSSR.

Chertok, B. E. (1996). *Rakety I lyidi* (448 s). Moskva: Kniga 2.

Chertok, B. (2006). *Rockets and people, volume 2: Creating a rocket industry* (669 p). Washington, DC: NASA.

Chertok, B. E. (2011, sentyabrya 23). Pora prosoedinit Lunu k Zemle. *Novaya gazeta*, 106.

Kopal, Z., & Carder, R. W. (1974). *Mapping of the moon. Past and present. Astrophysics and space science library*. Dordrecht: Reidel Publishing Company.

Linder, I. (1961). Simpozium "Luna". *Nayka i zhizn, 3*, 16–32.

Markov, A. V. (1963). *Veroyatnaya struktura I priroda obrazovaniy obratnoy storony Luny. V kn. Novoe o Lune* (s. 28–32). Izdatelstvo AN SSSR.

Mikhailov, A. A., & Straut, E. K. (1968). *Issledovaniya Luny. V kn. Uspehi SSSR v issledovanii kosmicheskogo prostranstva 1957–1967* (s. 250–268). Moskva: Isd, Nauka.

Pervye fotografii obratnoy storony Luny. (1959). (38 s). Izdatelstvo Akademii nauk SSSR. Moskva.

Shevchenko, V.V. (1986). *Issledovaniya obratnoy storony Luny. V kn. Problemy kompleksnogo issledovaniya Luny* (S. 3–16). Izdatelstvo Moskovskogo universiteta.

Whitaker, E. A. (1963). Evaluation of the soviet photographs of the moon's far side. In G. P. Kuiper & B. Middlehurts (Eds.), *The moon meteorites and comets* (p. 123). Chicago: The University of Chicago Press.

Wilkins, P., & Moore, P. (1955). *The moon*. London: Faber and Faber Limited. 388 p.

Chapter 2
Complete Maps of the Moon, Atlases and Globes

Abstract A preliminary review of the entire surface of the lunar sphere was completed in 1965 using images from the soviet spacecraft *Luna 3* and *Zond 3*. The main results were generalised in Part 2 of the *Atlas of the Lunar Farside*. The first *Complete Map of the Moon* at 1:5,000,000 scale was published in 1967. A second, improved edition of the *Complete Map of the Moon* at 1:5,000,000 scale was issued in 1969, including images taken by *Lunar Orbiter*. The images from *Zond 3, 6, 7, 8* and *Apollo 8, 11, 13* were coregistered by V. I. Chikmachev using the unified multiplex apparatus at GAISh. On a new complete map issued in 1979, more than 99.5 % of the lunar surface was represented. The illustrated relief was also published in the USSR on globes at 1:10,000,000 scale in 1967, 1969, 1974, 1979, 1984, 1989 and 1993. The methods developed for analysing images from *Luna 3* and *Zond 3, 6, 7, 8* proved to be very useful in the study of other planets.

First Complete Map of the Moon and the Second Part of the Atlas of the Lunar Farside (1967)

Between 1963 and 1965, five spacecraft—*Luna 4, 5, 6, 7, 8*—were launched using new carrier rockets with the aim of further research of the lunar trajectory and solving the problem of making a soft landing of apparatus on the lunar surface. However, Mstislav Vsevoldovich Keldysh insisted on obtaining high quality images of the lunar farside with low incidence angle illumination. This is how Boris Chertok writes about it in *Rockets and People*:

> "But Keldysh believes", he added, "that science will not forgive us if we pass up the opportunity to take better pictures with the Sun illuminating the Moon at an angle, when there will be great contrast between the shadows and light". Now we were finding possibilities for our *Semyorka* that we had never even thought of during its initial development. By building a third and then even a fourth stage onto the two stage military rocket, we were making the *Semyorka* into a launch vehicle for spacecraft to do fundamental research on the solar system. "It's difficult to argue with Keldysh", continued Korolev. "He's vice president of the Academy, I'm an academician, and we should enrich science with really fundamental discoveries, especially if they fall right in our laps".

V. Shevchenko et al., *Lunar and Planetary Cartography in Russia*, Astrophysics and Space Science Library 425, DOI 10.1007/978-3-319-21039-1_2

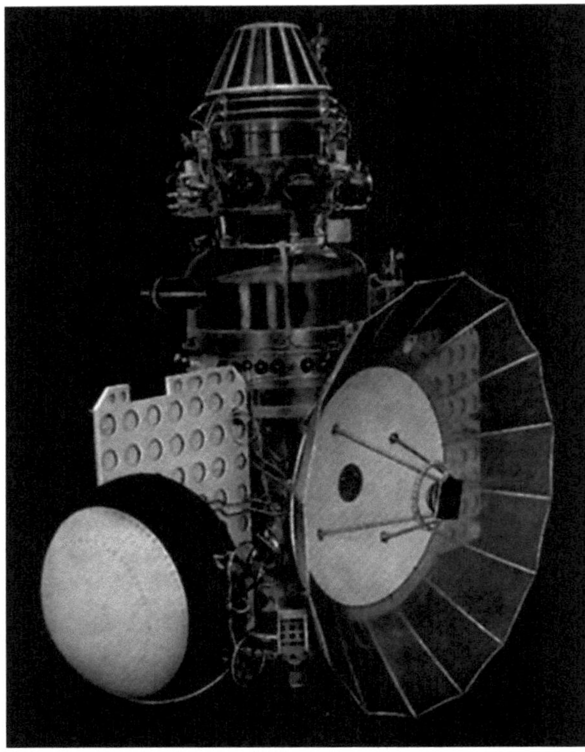

Fig. 2.1 Interplanetary spacecraft, *Zond 3*

This photography of the lunar farside was carried out by the *Zond 3* spacecraft, launched on 18th July 1965 (Fig. 2.1).

A fragment of one of the images of the eastern lunar farside, obtained by *Zond 3* with good illumination conditions for identifying relief, is shown in Fig. 2.2.

In 1964, Korolev founded the Department of Lunar and Planetary Physics at the Sternberg Astronomical Institute of Moscow State University. The department was headed by Dr. Yu. N. Lipskiy, who had earlier developed the techniques for processing the images from *Luna 3*. It was proposed that the new department should carry out research based on spacecraft images, covering primarily selenodesic and selenographic work. In this connection, a team of young specialists were recruited for this new field of Soviet science (Brian Harvey 2007). The first recruit was the graduate of MIIGAiK, V. V. Shevchenko, who already had some experience relating to the Moon, with results in several publications (Shevchenko 1963a, b; Lisina and Shevchenko 1965). Not long before, Shevchenko had given a talk at a student conference at GAISh, which was chaired by E. P. Aksenov, the future director of the institute. The talk was noticed by Professor N. P. Grushinskiy, who later recommended that Lipskiy employ the young scientist to work on lunar topics.

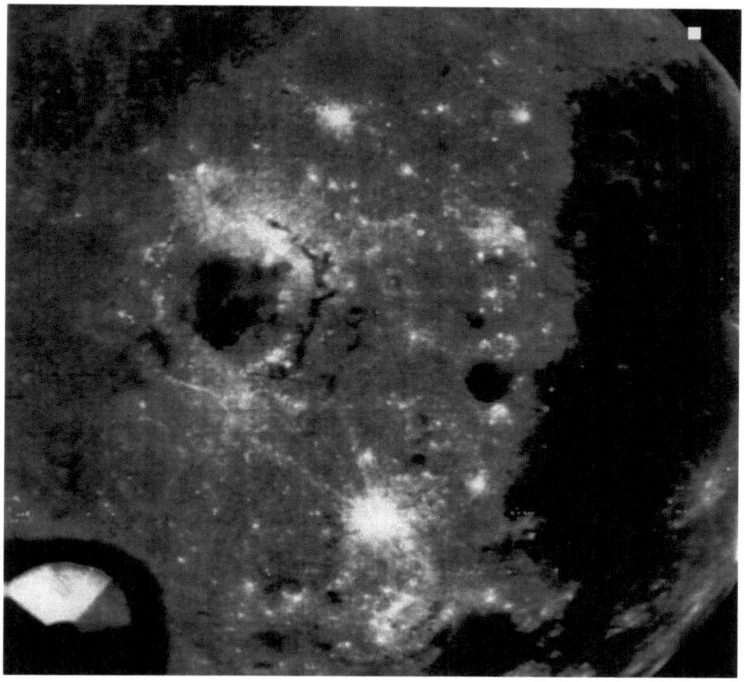

Fig. 2.2 Fragment of lunar surface image taken by *Zond 3*. The ring structure at the centre of the frame was subsequently named Hertzsprung, and the binary crater on its west edge was named Vavilov. The dark feature in the lower right of the image is a part of the lava flooding of Mare Orientale, not visible from Earth. The ring structure on the left side of the image near the terminator was named Korolev Basin (http://selena.sai.msu.ru/Home/Spacecrafts/Zond-3/Zond-3.htm)

Soon after, the department was filled out with members of the Laboratory of Photometry and Spectroscopy, who had taken part in the analysis of the first lunar farside images with Lipskiy: L. N. Bondarenko, K. I. Dekhtyareva, V. V. Novikov, Yu. P. Pskovskiy and M. M. Pospergelis. Later on, the department was joined by Zh. F. Rodionova, V. I. Chikmachev and K. B. Shingareva, graduates of MIIGAiK. These graduates of MIIGAiK were recruited in GAISh at Academician Korolev's request. We retain a copy of the letter, dated 16th November 1965, from Academician Korolev to the Rector of MIIGAiK, V. D. Bolshakov, requesting the placement of Zh. F. Rodionova and V. I. Chikmachev into organisation No. 651 with the next transfer to GAISh MGU. The Special Design Office (OKB-1) of the Central Design Office for Experimental Engineering of the USSR Ministry of Engineering, led by Korolev, fell under the same organisation code (Fig. 2.3).

Zh. F. Rodionova spent her pre-diploma internship at GAISh in the summer of 1965, working on the cartographic interpretation of the *Zond 3* images of the lunar farside (Fig. 2.4). The photography of parts of the nearside and eastern farside was

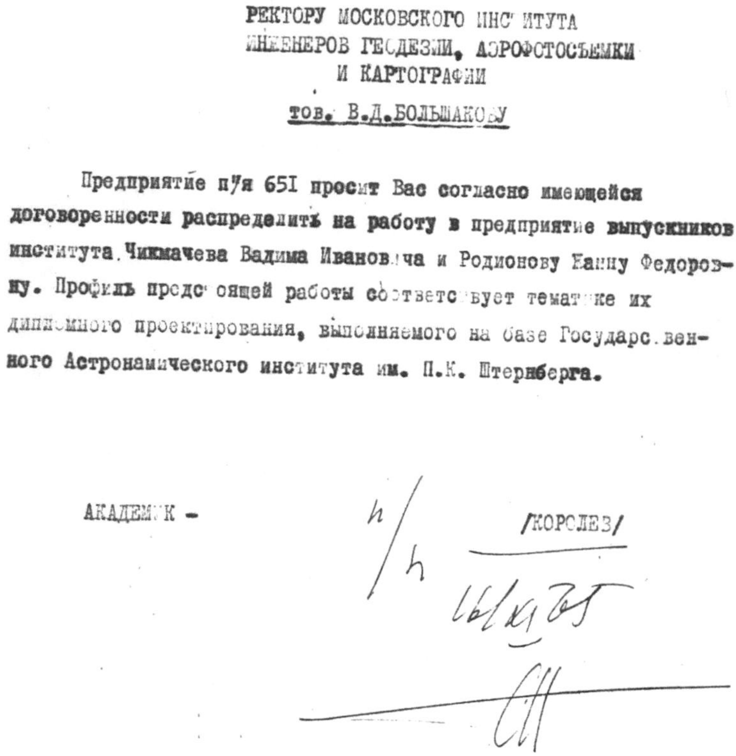

РЕКТОРУ МОСКОВСКОГО ИНСТИТУТА
ИНЖЕНЕРОВ ГЕОДЕЗИИ, АЭРОФОТОСЪЕМКИ
И КАРТОГРАФИИ
тов. В.Д.БОЛЬШАКОВУ

Предприятие п/я 651 просит Вас согласно имеющейся договоренности распределить на работу в предприятие выпускников института Чикмачева Вадима Ивановича и Родионову Жанну Федоровну. Профиль предстоящей работы соответствует тематике их дипломного проектирования, выполняемого на базе Государственного Астрономического института им. П.К. Штернберга.

АКАДЕМИК - /КОРОЛЕВ/

16/XI 65

Fig. 2.3 Copy of the letter from Korolev to the Rector of MIIGAiK, B. D. Bolshakov. Translation of the letter: "To V. D. Bol'shakov, Rector of the Moscow Institute of Geodesy, Aerial Photography and Cartography. Enterprise POB 651 requests that, according to the existing agreement, you assign Vadim Ivanovich Chikmachev and Zhanna Fyodorovna Rodionova to work in the enterprise. The profile of the upcoming work corresponds to the subject of their diploma projects, carried out at the Sternberg Astronomical Institute. *Academic Korolev 16.11.1965*."

carried out over 68 minutes from an altitude of 11.6 km. After the film was developed on board the spacecraft, the images were transmitted to Earth on 29th July from a distance of 2.2 million km. A repeat transmission was made from a distance of 30 million km. The data were processed as they were received. Lipskiy and his colleagues, Pskovskiy, Shevchenko, Pospergelis and A. A. Gurshstein, who had been temporarily seconded by Korolev from OKB-1 were in the office next to the Chief Engineer's. At that time, it was M. K. Tikhonravov's office, one of the pioneers of rocket technology. The Chief Engineer came in several times a day so that Lipskiy could report their progress.

Thus, in 1965, a preliminary review of the entire surface of the lunar sphere was completed. It became possible to construct a complete map and globe of the Moon, encompassing most of the lunar surface, including both the near and far hemispheres. The task was recognised as being of the highest scientific importance, and

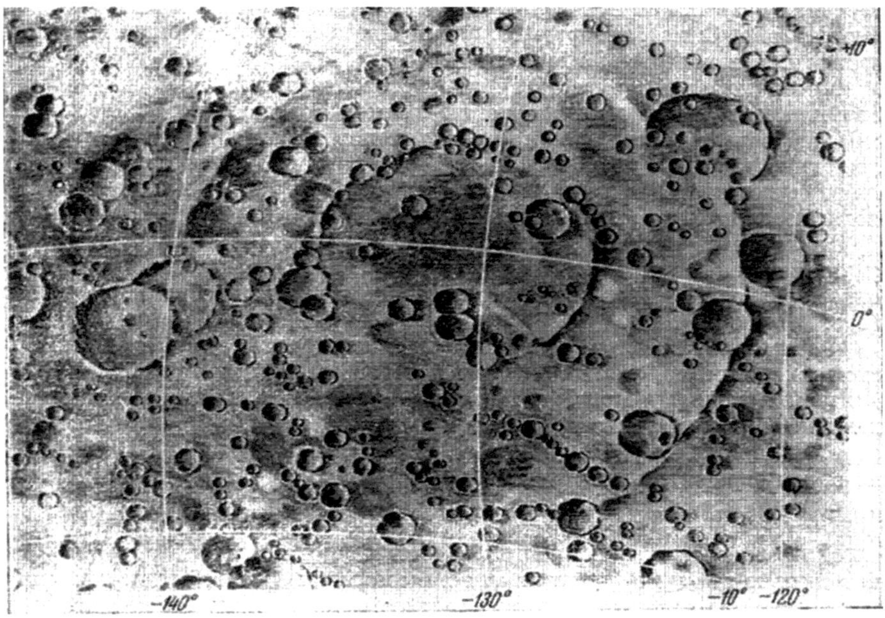

Fig. 2.4 Ring structure which was later named Hertzsprung. The double crater to the left of the ring was named Vavilov. Schematic map prepared by Zh. F. Rodionova

by a special resolution of the government of the USSR, Lipskiy's group was awarded a substantial government grant to achieve it.

Yury Naumovich Lipskiy was chosen, not by chance, to lead the work. From the time of his processing of the first farside images, he had won the respect and trust of Keldysh and Korolev. As well as his method from removing radio noise, there were several instances in his creative life when his scientific research significantly pushed forward the boundaries of knowledge, and his results remained hard to understand even for specialists.

In 1948 he defended his Ph.D. thesis entitled *Evaluation of the mass of the lunar atmosphere by polarisation studies of its surface*. The work was completed under the supervision of Academician V. G. Fesenkov, one of the founders of modern astrophysics, who obtained his doctoral degree at the Sorbonne. A difficulty was that the level of knowledge of the physics of the lunar exosphere was very low at the time. It was considered obvious that the exosphere could be made up only of gas components at very low density, assuming a near total absence of atmosphere. For this reason, when Lipskiy's pioneering, high-precision polarisation observations identified increased illumination in the region of the cusps of the nearside lunar disc, this effect—later known as the Lipskiy effect—could be understood by the majority of astronomers only as the uncertain indication of a gas layer: of an atmosphere, which, as everyone knows, is not possible on the Moon. Only a few of Lipskiy's colleagues—Fesenkov, in particular—understood that a discovery had been made, the physics of which had still to be explained. And only in our time, thanks to space research, has it been proved that the glow in the lunar exosphere is a real phenomenon which shows the presence of another component in the space around the Moon: a significant mass of fine cosmic dust. Typical observations using telescopes of the 1940s and 1950s could not reveal the effect. Only Lipskiy's high-sensitivity polarisation experiments were up to the task, and he was, in this respect, well ahead of his time. Today we know, thanks to the results of the LEAM instrument (Lunar Ejecta and Meteorites) which operated during the *Apollo 17* mission, as well as limb observations carried out by ESA's *SMART-1* and NASA's *Clementine* spacecraft, that Lipskiy's observations were an outstanding achievement of his time.

The first sketches of fragments of the Complete Map of the Moon were made by Zh. F. Rodionova at the end of 1965 (Fig. 2.4). Several were published in an article (Lipskiy et al. 1966) and in the Atlas of the Lunar Farside, Part 2 (Fig. 2.5).

When the material for the construction of the map were being prepared in the Department of Lunar and Planetary Physics at GAISh, specialists from the USSR Topogeodesic Service also worked on the construction of the nine sheets of the Complete Map of the Moon at 1: 5,000,000 scale and the segments of the Lunar Globe at 1:10,000,000 scale.

The nomenclature for lunar features had to be significantly expanded to be able to denote the newly identified structures on the surface. Proposals for names for the structures were made by a commission of the Presidium of the USSR Academy of Sciences under the chairmanship of Academician Keldysh. The technical preparation of the new names for the complete lunar map and globe was done by K. B. Shingareva of the Department of Lunar and Planetary Physics and the linguistic expert N. B. Lavrova, both at GAISh (Shingareva and Burba 1977).

The main results of the preliminary review of the entire surface of the Moon, made on the basis of *Luna 3* (Atlas Obratnoy Storonu Lunu 1960) and *Zond 3* photographs, were generalised in Part 2 of the Atlas of the Lunar Farside under Lipskiy's guidance. Colleagues from GAISh (the lead organisation); the USSR Academy of Sciences; Pulkovo Observatory; Goloseevo Astronomical Observatory (GAO); Kharkov University Observatory; USSR Topogeodesic Service; Academy of Signals; and Central Institute for Geodesy, Aerial Photography and Cartography all took part in the preparation of different sections of the atlas, which was published in 1967 (Atlas of the Lunar Farside, Part 2 1967). Using the images from *Zond 3*, the authors made a detailed study of the lunar farside. The data included analyses of the density distributions, sizes and forms of craters of the eastern sector of the farside, as well as photometric properties of selected features of the sector, and topographic characteristics of other features determined by photometric methods. The appendix includes four lists of proposed names covering 233 features on both the near and far sides of the Moon with biographical information on the prominent scientists and engineers after whom they were given.

An important section of the atlas covered the extension of the selenographic coordinate system to the farside, which was carried out in the Department of Lunar and Planetary Physics. This development provided the coordinate basis for the first Complete Map of the Moon. Figures 2.6 and 2.7 show sheets 1 and 2 of the map

Fig. 2.5 The three parts of the Atlas of the Lunar Farside, published in 1960, 1967, and 1975 in the USSR (authors' photo)

(Complete Map of the Moon, 1967). The atlas presents photomaps of the farside with the selenographic coordinate grid and the methods used for tying the images. The catalogue of control points for the eastern sector included 59 features. For the nearside, the control point network was based on the GAO catalogue (500 points), and the Arthur catalogue (Arthur et al. 1963, 1964, 1965, 1966) was used for the edge zones.

The first Complete Map of the Moon at a scale of 1:5,000,000 was compiled under Lipskiy's guidance on the basis of *Luna 3* and *Zond 3* images and published in 1967. The map covers 95 % on the lunar surface. The regions between ±60° latitude are constructed in arbitrary cylindrical projection, and the circumpolar regions are represented in azimuthal projection at scale 1:10,000,000. The half scale is explained by the relative lack of coverage of high latitudes at the time. The map consists of nine sheets: six for the equatorial region out to ±60° latitude and three for the polar regions and the list of names. In the arbitrary cylindrical projection, the angular distortions do not exceed ±5° up to ±50° latitude. At 60° the angular distortion is below 14°, but the area is distorted very much less than, for example, in the Mercator projection. The standard parallels, for which the scale remains true, are taken to be ±30°. In the direct conformal azimuthal projection, used for the polar regions, the standard parallels are ±70°. The sheets were designed so that the nearside hemisphere lies at the centre of the map, and the farside is divided into east and west sectors. The original division of the map was intended to be convenient for working with individual sheets as well as with the map as a whole on the wall (Figs. 2.6 and 2.7).

The successful imaging of the lunar farside motivated research of the lunar surface, both using ground-based observations and space observations in the USA. Between 1960 and 1963 the *Photographic Atlas of the Moon*, edited by G. Kuiper,

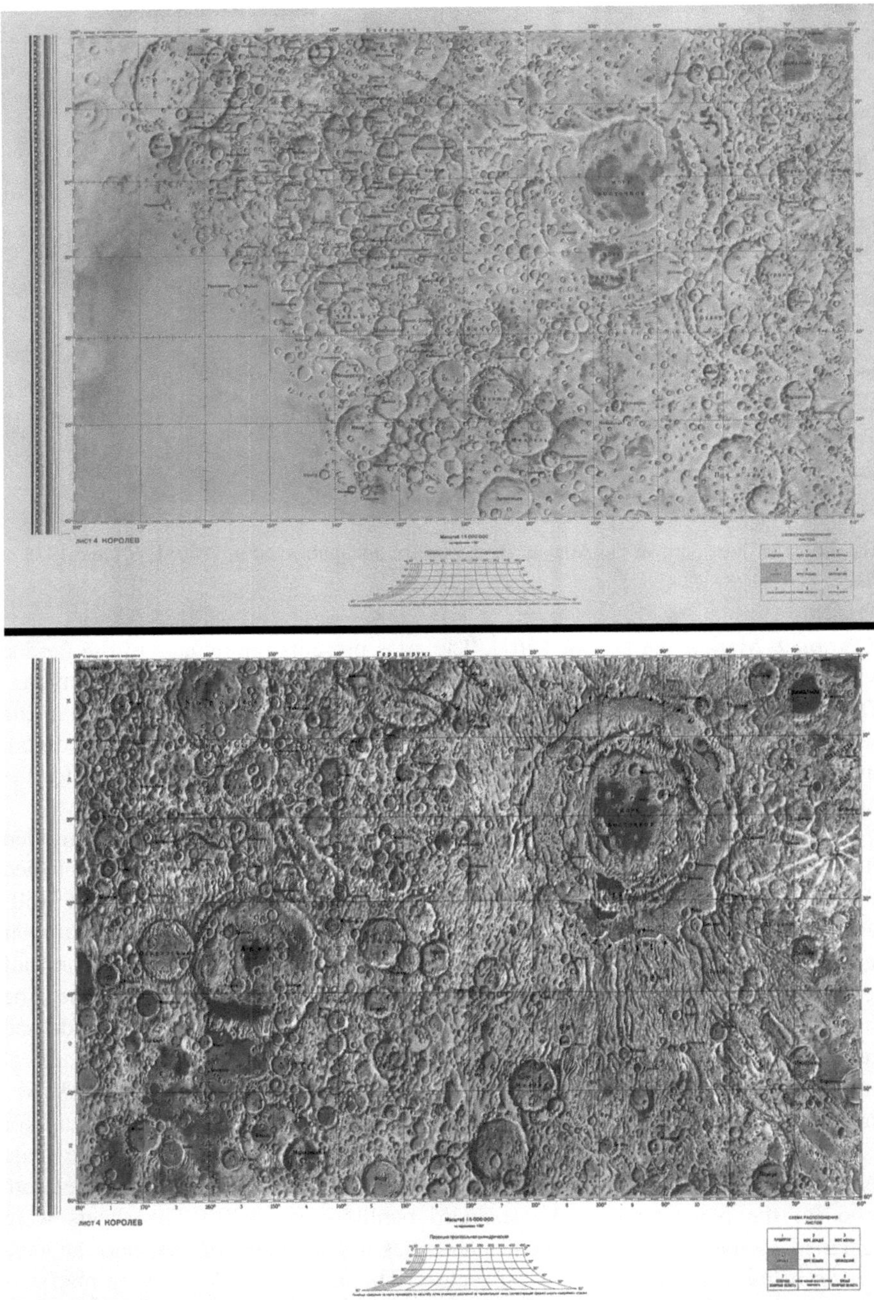

Fig. 2.6 Fragment of the *Complete Map of the Moon,* sheet 4 (1967, 1979). The half-tone image of the lunar surface was produced by cartographic illustrator V. P. Savchenko (hand tinted)

Fig. 2.7 Fragment of the *Complete Map of the Moon,* sheet 2 (1967). The half-tone image of the lunar surface was produced by cartographic illustrator V. V. Sokolov (hand tinted)

the *Orthographic Atlas of the Moon* and *Rectified Atlas of the Moon* had appeared in the USA and served as the basis for many maps of the nearside. In 1965, the *Photographic Atlas of the Moon* (by Kopal, Klepesta and Rackham), was published, consisting of Earth-based images from a 61-inch telescope at Pic du Midi in France, in collaboration with American specialists. These data and maps of the nearside hemisphere were later used in the compilation of maps at GAISh and associated institutions for increasing the precision and detail of illustration.

Complete Map of the Moon (2nd Edition, 1969)

A second, improved edition of the *Complete Map of the Moon* was issued in 1969. It included images taken by *Lunar Orbiter*, and the polar regions were now shown at 1:5,000,000 scale as a result of the availability of polar images with greater detail. Since the set of control points derived from the *Zond 3* images was tied to the nearside coordinate system only from the west, a systematic accumulation of position errors was discovered on the 1967 map, increasing in the south-west direction (Fig. 2.8).

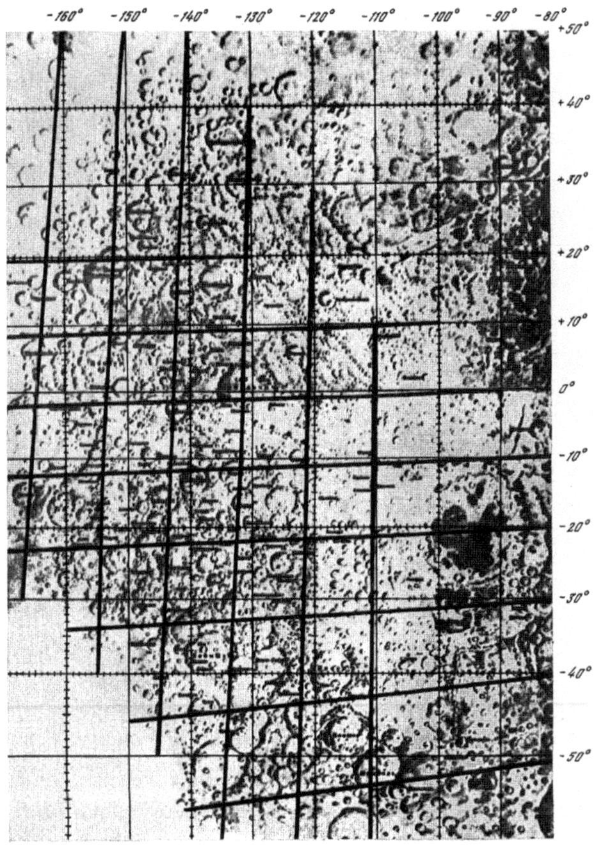

Fig. 2.8 Fragment of the first *Complete Map of the Moon,* where the *heavy lines* indicate the coordinate grid determined by the method of rectification to orienting points (Lipskiy and Chikmachev 1969). The discovered offset between grids was corrected in the 2nd edition

An improved coordinate grid in the eastern sector of the farside (Fig. 2.9) was obtained by the use of a more sophisticated means of tying the images together using rectification to orienting points (Lipskiy and Chikmachev 1969).

Preliminary drafts, showing the borders of the lunar maria, ray systems, and relative highs and lows of the lunar relief, were made at GAISh to be used by the cartographic illustrators of the Topogeodesic Service during their half-tone illustration of the surface. The models were made on blue prints of map sheets with the original mounted photographs. The boundaries of maria and ray systems on the drafts were shown using full Moon images of the near and far hemispheres. To compile the drafts of relative heights for the nearside, topographic maps from various publications and descriptions of the lunar surface were used. Images from the farside were tied into the selenographic coordinate system by reprojection onto a spherical screen (Lipskiy and Chikmachev 1969). Images from *Zond 6* and *Lunar Orbiter 1* which included the Earth in the frame were used as an independent control of the planned positioning of features (Shevchenko 1968).

Fig. 2.9 *Zond 3* image with corrected coordinate grid overlaid

An apparatus was constructed in the Department of Lunar and Planetary Physics (GAISh) in collaboration with NPO Energomash, which allowed spacecraft images to be projected onto a precision spherical screen which was marked with a grid of meridians and parallels, in order to obtain the view from an arbitrary observation point. This technique was applied extensively to produce a new generation of lunar maps and globes from new images of both the near and far sides (Fig. 2.10).

Spacecraft flights around the Moon and returning to Earth enabled photographs of both the Moon and Earth to be obtained with significantly higher quality than transmitted images. *Zond 5* flew around the Moon at a distance of 1950 km on 18th September 1968, and *Zond 6* photographed the Moon from 2420 km on 12th November 1968. However, on 27th December 1968, the American manned spacecraft *Apollo 8* successfully flew around the Moon and returned Frank Borman, James Lowell and William Anders to Earth. An important new page had been opened in space exploration.

Zond 7 and *Zond 8* also returned photographic films to Earth, including images of nearly the whole of the western hemisphere at relatively high resolution. The

Fig. 2.10 Zh. F. Rodionova and V. I Chikmachev in the Department of Lunar and Planetary Physics (GAISh) working with the unified multiplex apparatus, equipped with a precision spherical screen for tying the coordinate systems of spacecraft images

processing of these photomaterials on Earth normally exceeded the quality of transmitted images. A series of maps of the lunar farside from original photographic materials from *Zond 6, 8* was published by MIIGAiK under the guidance of V. D. Bolshakov.

The images from *Zond 3, 6, 7, 8* and *Apollo 8, 11, 13* were tied together by Chikmachev using the unified multiplex apparatus at GAISh which allowed, by optical means, the reconstruction of the projection of the light rays at the moment the image was taken. The apparatus likewise allowed the correction of inclined or rotated images. Figure 2.11 shows an example of an image returned by *Zond 8* processed this way.

Thus, the "white patch" at the south pole was reduced in size: on the new map, more than 99.5 % of the lunar surface was represented. The reflective

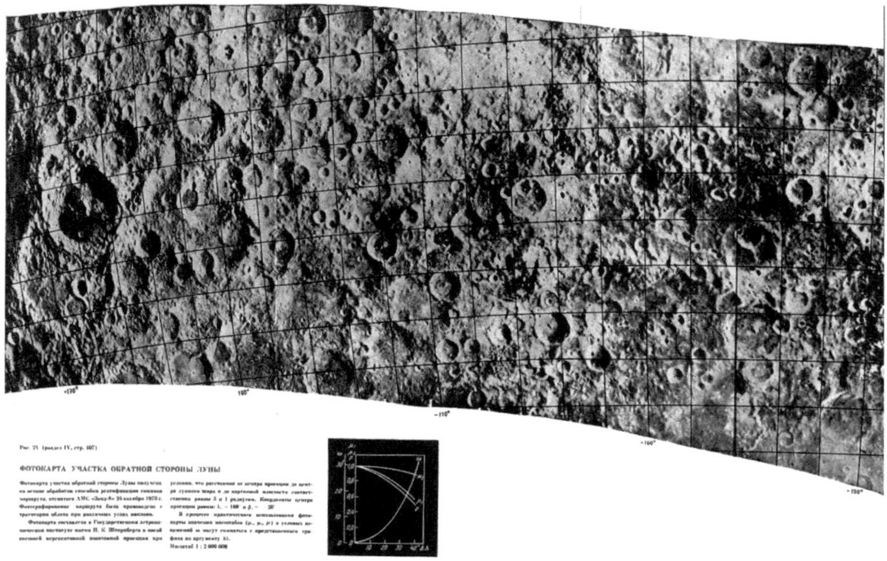

Fig. 2.11 One of a series of images obtained by *Zond 8* and returned to Earth. The coordinate grid was added using the unified multiplex

characteristics of surface structures were represented by variations of tone in the colour washes (Lipskiy and Rodionova 1978).

Table 2.1 presents the sequential image numbers and the number of control points from the nearside used to tie the frame into the selenographic coordinate system.

For the annotation of craters on the 1969 *Complete Map of the Moon*, seven gradations of text size were used: from 3.0 mm for large basins down to 2.0 mm for craters less than 15 km diameter. Maria, depending on their size and characteristics, were indicated with italic text sizes from 5 mm down to 1.5 mm. Mountain chains were indicated in four text sizes from 3.0 down to 1.0 mm. Capes and peaks were marked with 1.5 mm lettering, valleys from 2.0 down to 1.5 mm, fissures and grooves with letters 1.5 down to 1.2 mm depending on their extent. The scheme for the lettering was designed at TsNIIGAiK (Lipskiy et al. 1977).

Atlas of the Lunar Farside (1975) and Complete Map of the Moon (3rd Edition, 1979)

The third, significantly extended edition of the *Complete Map of the Moon* was published in 1979 (Polnaya Karta Lunu 1:5 000 000 1979). Particular attention was given to the illustration of the relief of both the near and far sides, to the representation of maria, craters with flood floors, with ray systems, and to crater chains. Care

Table 2.1 Parameters of the spacecraft images used for coordinate registration of details on the lunar farside. Column labels: Spacecraft; Image number; Focal length, mm; Frame size, cm; Spacecraft altitude, km; No. of control points on lunar nearside. Spacecraft names: Zond 6, 8 (Western hemisphere); Apollo 8, 11, 13 (Eastern hemisphere)

Космический аппарат	№ снимка	Фокусное расстоЯние обзектиеа, мм	Размер кад ра, см	Вьlсота сземки, км	Число опорньlх точек видимого полуwа риЯ Луньl
Западное полушарие					
«Зонд-6»	9	400	13 × 18	8870	65
	10	400	13 × 18	8860	67
	11	400	13 × 18	8850	37
	12	400	13 × 18	8840	22
	13	400	13 × 18	8820	2
	45	400	13 × 18	8430	–
	50	400	13 × 18	8370	–
	53	400	13 × 18	8330	–
«Зонд-8»	4	400	13 × 18	9613	138
	5	400	13 × 18	9592	143
	37	400	13 × 18	8906	112
	38	400	13 × 18	8879	115
Восточное полушарие					
«Аполлон-8»	2484	80	5,5 × 5,5	2920	25
	2485	80	5,5 × 5,5	3230	36
	2506	80	5,5 × 5,5	5050	37
«Аполлон-11»	6665	80	5,5 × 5,5	4950	75
«Аполлон-13»	8765	60	5,5 × 5,5	3740	30
	8795	60	5,5 × 5,5	5780	26

was given, for example, to accurately portray shallow craters or craters with low rims in relation to their depth or the steep slopes at the interior of the rim by comparison with the exterior slope. The colour range used in this edition was chosen taking account of the natural colour of the Moon with the aim to achieve the highest fidelity in reproduction of the relief. The catalogue of named features in Russian and Latin, compiled at GAISh, is shown on sheets 7, 8 and 9 of the map. The map was intended for scientific research applications as well as for broad circulation to those interested in the Moon.

The navigational projection of the map permitted its use for overlaying the trajectories of lunar spacecraft and the design of space missions. It was used as a basis for the overlay of various types of information describing the properties of the surface or the geological structure (Rodionova 1983). It was used to study the distributions of lunar features and their relationships and to measure the areas of maria or large craters. A *Morphological Catalogue of Lunar Craters* (Rodionova et al. 1987) was compiled using the map. Table 2.2 presents the maria area measurements by Rodionova (1969, 1972) and Westfall (1970).

Table 2.2 Measurements of surface area of lunar maria. Column labels: Russian name, Latin name, Area, thousand km^2 (Rodionova 1968, 1972; Westfall 1970

Название	Т ранслитерация	Площадь, тыс. км^2 Ж. Ф. Родионова, 1968 г.	Дж. Вестфэл,[e*] 1970 г.	Ж. Ф. Родионова, 1971 г.
1	2	3	4	5
Видимая сторона				
Океан Бурь[a*]	Oceanus Procellarum	2105	2147	2102
Море Дождей	Mare Imbrium	830	835	829
Море Холода[b*]	Mare Frigoris	347	433	436
Море Спокойствия	Mare Tranquillitatis	430	408	421
Море Изобилия	Mare Fecunditatis	311	334	326
Море Ясности	Mare Serenetatis	305	312	303
Море Облаков	Mare Nubium	253	240	254
Море Кризисов[c*]	Mare Crisium	180	197	176
Море Влажности	Mare Humorum	113	115	113
Море Смита	Mare Smythii	109	77	104
Залив Росы[d*]	Sinus Roris	107	291	–
Море Нектара	Mare Nectaris	96	99	101
Море Познанное[d*]	Mare Gognitum	73	–	–
Озеро Сновидений	Lacus Somniorum	66	–	72
Море Краевое	Mare Marginis	64	82	62
Море Паров	Mare Vaporum	51	53	55
Залив Дентральный	Sinus Medii	48	49	52
Залив Зноя	Sinus Aestuum	–	43	40
Залив Радуги	Sinus Iridum	–	–	39
Еолото Эпидемий	Palus Epidemiarum	40	29	27
Море Гумбольдта	Mare Humboldtianum	23	28	22
Море Волн	Mare Undarum	13	19	21
Море Гниения	Palus Putredinis	–	14	12

(continued)

Table 2.2 (continued)

Название	Т ранслитерация	Площадь, тыс. км^2		
		Ж. Ф. Родионова, 1968 г.	Дж. Вестфэл,[e*] 1970 г.	Ж. Ф. Родионова, 1971 г.
1	2	3	4	5
Море Пены	Mare Spumans	14	14	16
Море Весны	Mare Veris	20	16	12
Озеро Смерти	Lacus Mortis		14	12
Море змеи	Mare Anguis	8	–	10
Море Струве	Mare Struve	–	–	4
Море Осени	Mare Autumni	–	4	3
Море Лета	Mare Aestatis	4	4	1
Обратная сторона				
Море Южное	Mare Australe	147	148	151
Море Восточное	Mare Orientale	65	60	54
Море Москвы	Mare Moscoviense	67	49	50
Море Мечты	Mare Ingenii	–	27	15
Море Мирное	Mare Pasificus	20	–	13

[a*]В Океан Еурь включено Море Познанное и Залив Росы
[b*]В границы Моря Холода, в отличие от измерений 1968 г., включены Озеро Смерти и прилегающие к нему с востока морские поверхности в соответствии с картой [13]
[c*]В площадь Моря Кризисов, в отличие от ДЖ. Вестфэла, не включена площадь Моря Змеи
[d*]Площади Залива Росы и Моря Познанноого не измерены ввиду неоггределенности гранип
[e*]В графе данных Дж. Вестфэла приведены площади морских поверхностей без «островов». (Данные Вестфэла округлены до целых тысяч *км*^{2})

Using the original illustrated relief for the 3rd edition, a reduced version at 1:10,000,000 (Polnaya Karta Lunu 1:10 000 000 1979) scale was prepared for publication on a single sheet (Fig. 2.12). The idea for such a map was Lipskiy's, but sadly he did not live to see it published in 1979. Lipskiy's name can be found at the centre of the lunar farside on the crater named after him. The *Map of the Moon* includes Rodionova's maria measurements, a map of lunar rock types compiled by Shevchenko (Shevchenko 1977), data on the key stages of spacecraft exploration and the basic parameters of the Moon. A second edition of the map with additional information was published in 1985 (Karta Lunu 1:10 000 000 1985).

Figure 2.13 represents the region of Mare Moscoviense which was shown on the *Complete map of the Moon* issued in 1967, 1969, 1979.

The illustrated relief was also published in the USSR on globes at 1:10,000,000 scale in 1967, 1969, 1974, 1979, 1984, 1989 and 1993. For each edition there were

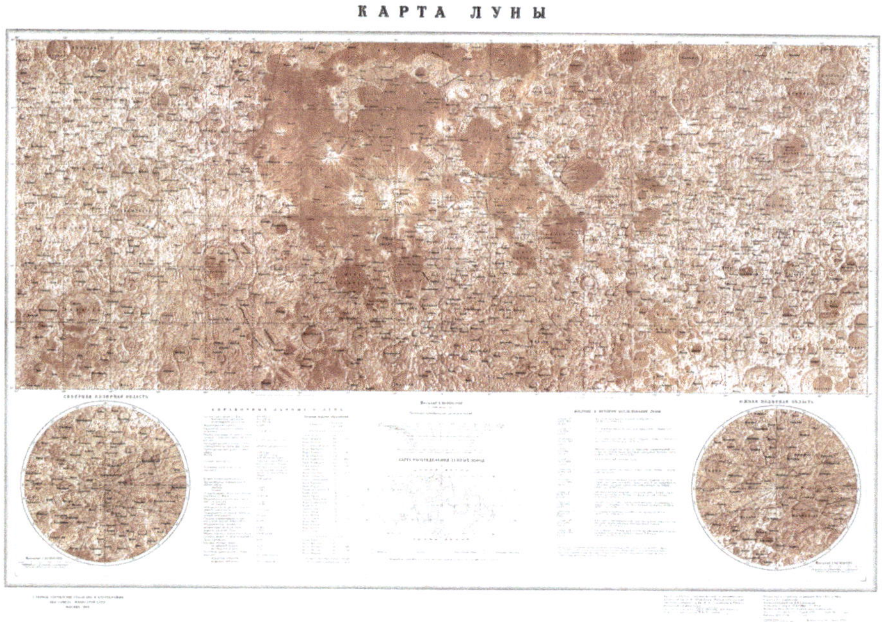

Fig. 2.12 *Map of the Moon* at 1:10,000,000 scale (1985)

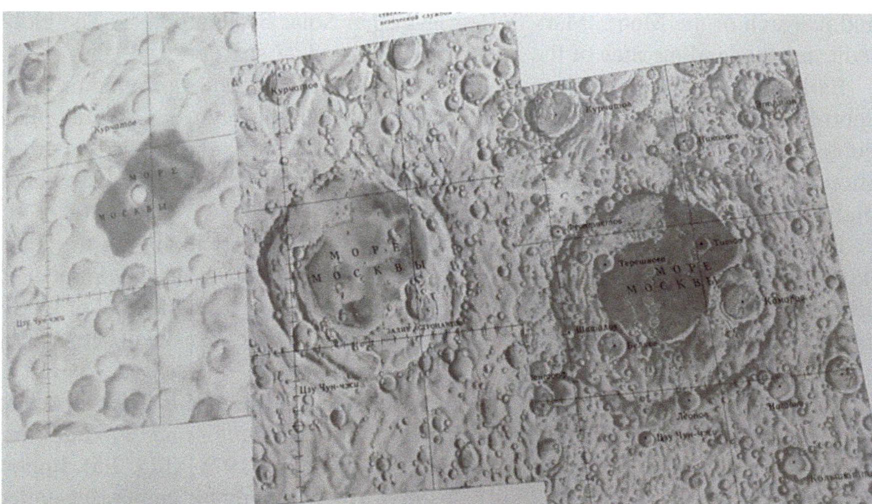

Fig. 2.13 Region of Mare Moscovience is shown on the *Complete map of the Moon* issued in 1967, 1969, 1979

changes made to the naming according to decisions of the International Astronomical Union. The originals for the lunar globes (Globus Lunu 1:10 000 000 1967, 1969, 1974, 1979) were prepared in the form of 12 segments, covering 30° in longitude and ±80° in latitude, together with two polar discs in azimuthal equidistant projection. The segments for the farside were prepared using an original method of transformation using a spherical screen. By special order in 1970, several globes were produced with a diameter of 1.2 m at a scale of 1:3,000,000.

New possibilities for producing planetary globes appeared with the application of thermoplastic materials, allowing a whole hemisphere to be made in one piece. A flat cartographic illustration of a hemisphere is produced in azimuthal projection, taking account of the deformation which occurs in forming the plane into a hemisphere (Boginskiy et al. 1990). A molding method was developed at TsNIIGAiK, whereby a plunger is heated to a certain temperature and pressed at constant speed into a thermoplastic disc, printed with one hemisphere and held by an annular clamp. Two hemispheres are made and glued together at the equator. PKO Kartografiya, GAISh and TsNIIGAiK collaborated to produce a 32 cm lunar globe, released in 1990 in two versions: with and without internal illumination.

It is worth noting in particular the highly skillful work of the cartographic illustrator, V. V. Sokolov, who produced several editions of the lunar map at GAISh at various scales, and original segments for many of the lunar globes (Fig. 2.14).

Figure 2.15 shows the lunar globe at the Lavochkin Museum, known for its exhibition of lunar and planetary spacecraft which carried out exploratory missions and research of the Moon, Mars, Venus and other Solar System bodies. The globe indicates the landing sites of the *Luna 16, 20, 21* and *24* spacecraft.

Early in 1968, Wernher von Braun, the head of the Apollo programme, had approached the USSR Academy of Sciences with a request for the set of lunar maps published by GAISh. They were sent, and Figure 2.16 shows the letter of appreciation he wrote. The exchange took place shortly before the realisation of manned flights to the Moon in the *Apollo* programme.

In many cases, the lunar cartographic materials—the first including the lunar farside—became a special kind of souvenir, given by the Soviet government to other states at the highest levels. Figure 2.17 shows a scene from the visit of USSR pilot-cosmonaut P. R. Popovich to Yugoslavia, when he gives President Tito a copy of the *Complete Map of the Moon.*

The results of the next phase of study of the lunar surface with a new series of spacecraft were presented in the third part of the *Atlas of the Lunar Farside*, first published in 1975 (Atlas Obratnoy Storonu Lunu. Chast 3 1975) (Fig. 2.5). Important results were obtained from *Zond 6, 7* and *8*. The photographic cameras installed on the spacecraft were prepared by the Moscow Institute of Geodesy, Aerial Photography and Cartography (MIIGAiK). The significant change was that the photographic film was returned to Earth, so that the development and processing could be carried out in normal laboratory conditions. The negatives had high photometric and interpretive quality. The absence of line and other defects, which

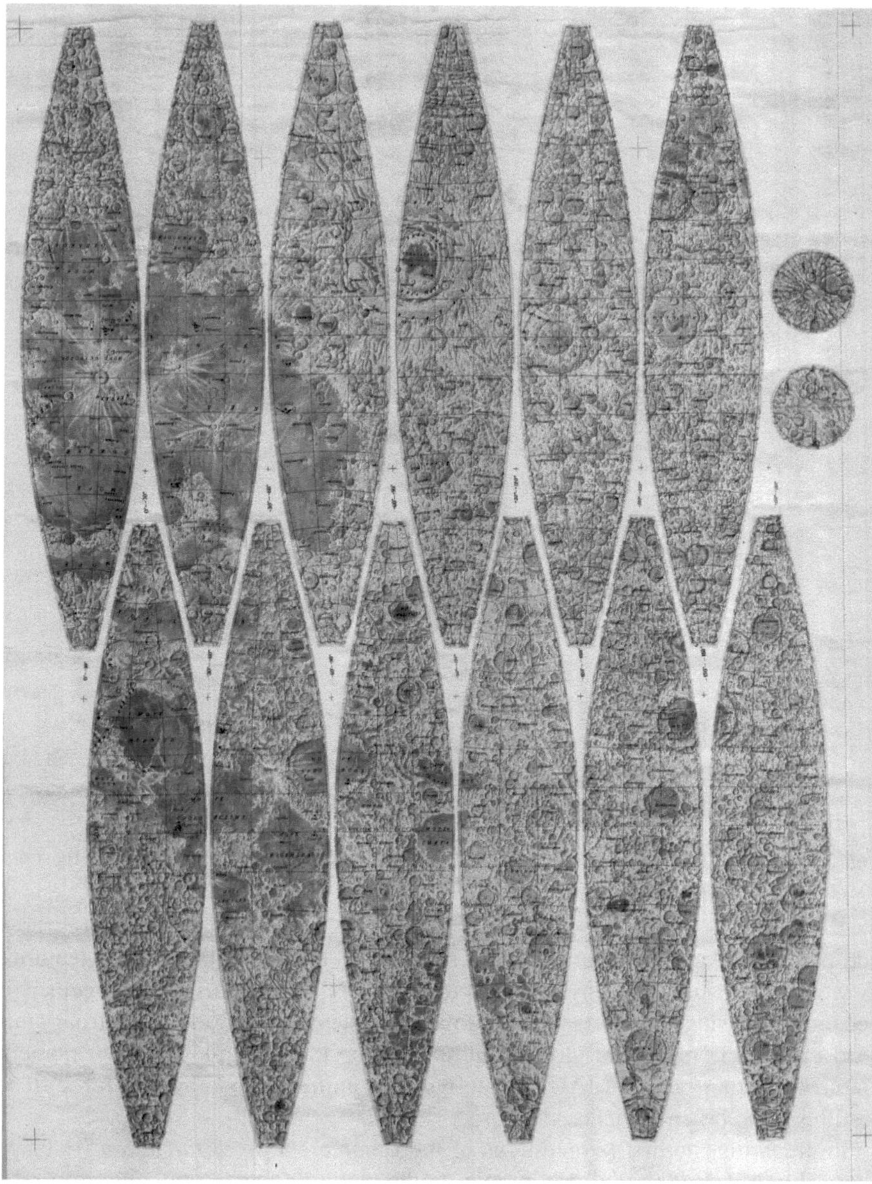

Fig. 2.14 Printed segments for the Lunar globe at scale 1:10,000,000 by cartographic illustrator, V. V. Sokolov (1979)

were previously introduced by the on board processing and transmission of the images to Earth greatly improved the information content of the materials.

As for the first two parts of the atlas, the study of the distant lunar hemisphere, not accessible to Earth-based telescopes, gave researchers a lot of new information,

Fig. 2.15 Lunar globe at the S. A. Lavochkin Museum showing the landing sites of the *Luna 16, 20, 21* and *24* spacecraft

and permitted work to begin on the morphological classification of lunar features, as well and to study the physical characteristics of both hemispheres together. It became increasingly important to improve the system of selenodesic control points used to compile maps and globes, and to develop the principles of lunar nomenclature (together with the IAU) and to define a unified albedo system for surface features in both hemispheres.

In preparing the first two editions of the *Complete Map of the Moon* (Polnaya Karta Lunu 1:5 000 000 1967, 1969), the control point catalogue from the Goloseevo Observatory was used. The ongoing selenodesic work at GAISh using this data and others, eventually revealed its deficiencies: both the control point precision and the point density decreased towards the edge of the nearside, making an effective boundary at ±70° latitude and longitude. There were no control points in the libration zones. An improved catalogue was introduced at GAISh which

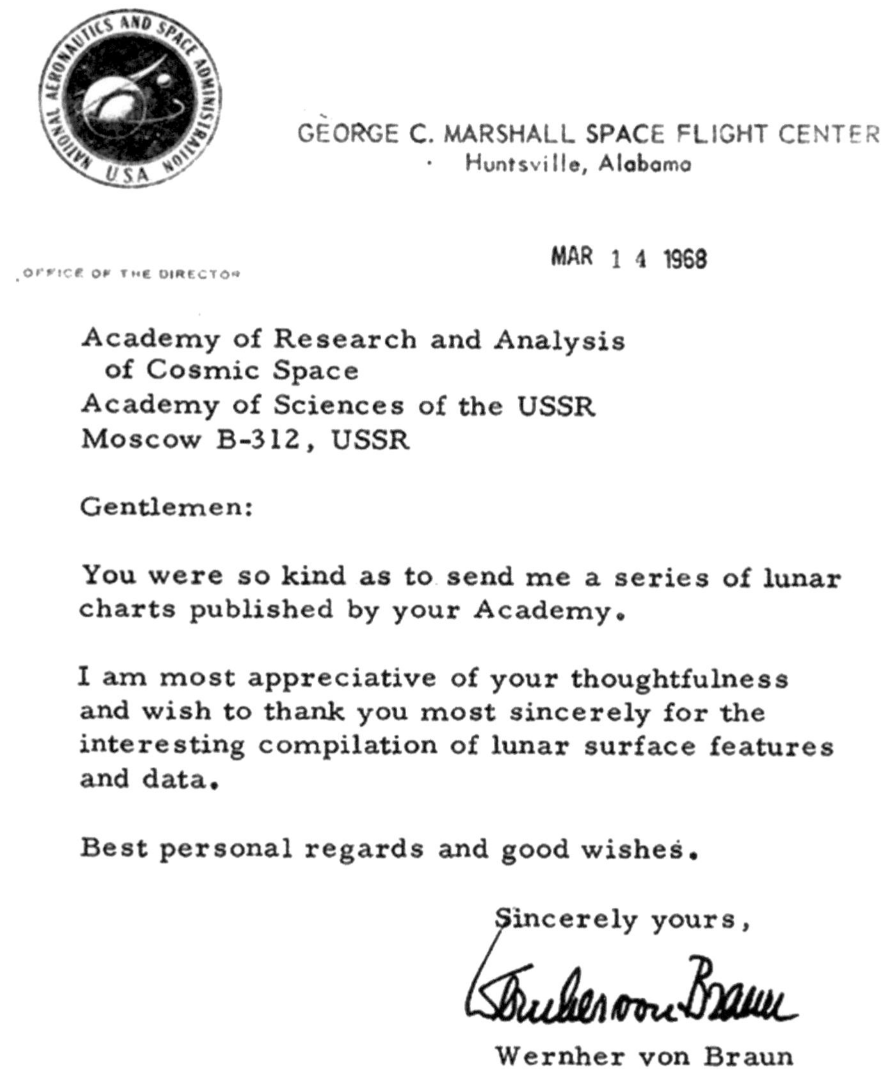

GEORGE C. MARSHALL SPACE FLIGHT CENTER
Huntsville, Alabama

OFFICE OF THE DIRECTOR

MAR 14 1968

Academy of Research and Analysis
of Cosmic Space
Academy of Sciences of the USSR
Moscow B-312, USSR

Gentlemen:

You were so kind as to send me a series of lunar charts published by your Academy.

I am most appreciative of your thoughtfulness and wish to thank you most sincerely for the interesting compilation of lunar surface features and data.

Best personal regards and good wishes.

Sincerely yours,

Wernher von Braun

Fig. 2.16 Letter of appreciation from Wernher von Braun to the USSR Academy of Sciences

brought all the data into a unified system (Lipskiy et al. 1973). This information was included in part 3 of the atlas, together with a description of the methods and apparatus used at GAISh, IKI (Institute of Space Research) and MIIGAiK. The unified multiplex apparatus for geometry correction was also described. While the unified multiplex apparatus was used at GAISh, the images from *Zond 6, 7, 8* were studied simultaneously at MIIGAiK and IKI. Using different approaches, it was possible to perfect the techniques for analytical trajectory phototriangulation and compare the precision of control point catalogues made by different methods.

Fig. 2.17 USSR pilot-cosmonaut P. R. Popovich presents a copy of the *Complete Map of the Moon* to President Tito of Yugoslavia during a visit to Belgrade in 1968 (collib.com)

A substantial section of part 3 of the atlas was devoted to the results of photometric analysis of images from the farside and the inclusion of these into a unified system with photometric measurements from the nearside (Shevchenko 1980).

Photometric and morphological study of the lunar surface in conjunction with selenodetic work was important. The later exploration of Mars, Venus and Mercury showed that the surfaces of these bodies, despite significant differences in many fundamental parameters, have a lot in common. The methods developed for analysing images from *Luna 3* and *Zond 3, 6, 7, 8* for the three parts of the *Atlas of the Farside of the Moon* and the complete map and globes proved to be very useful in the study of other planets. Work on the atlas was the first experience of rigorous extraction of the information contained in satellite images: photometric, morphological and selenodetic. Figure 2.18 shows the members of the Department of Lunar and Planetary Physics, who took part in the pioneering work in the first stage of studying the lunar farside.

Overview Maps of the Moon, Published at MIIGAiK

A distinctive feature of the site maps of the lunar surface, produced at MIIGAiK based on images from *Zond 6, 7, 8* is the use, along with relief shading, of line symbols reflecting the morphological structure and age characteristics of the

Fig. 2.18 Members of the Department of Lunar and Planetary Physics, who took part in the pioneering work in the first stage of studying the lunar farside. Clockwise: V. Nikonov, V. Titov, T. Skobeleva, K. Dekhtyareva, Zh. Rodionova, A. Sanovich, V. Shevchenko. In the background is the 3rd edition of the *Complete Map of the Moon* (1979)

surface topography. The techniques used to produce a map at scale 1: 2,000,000 of a portion of the equatorial belt of the farside based on *Zond 6* images was described by N. M. Volkov (1978). The map sheet, covering an area of $\pm 20°$ in latitude and 200°–270° longitude, was made in four versions. Variant A used a colour wash to represent the structure of the slopes, their steepness, fragmentation, degree of degradation and the sequence of formation. Variant B was a structural and morphological map constructed using line symbols that reflect three age systems: lower, middle and upper, as well as some intermediate divisions (Fig. 2.19). Variant V shows the age characteristics of the relief using a colour wash. Variant G is presented in the form of a topographic map which, in addition to containing the relief shown by colour wash, also shows characteristics of the surface as dashed lines. The 1:1,000,000 scale map (Karta Obratnoy Storonu Lunu 1977), produced from images from *Zond 8* in Mercator projection, consists of two sheets of the equatorial region. This map provided two variants of representation of the lunar

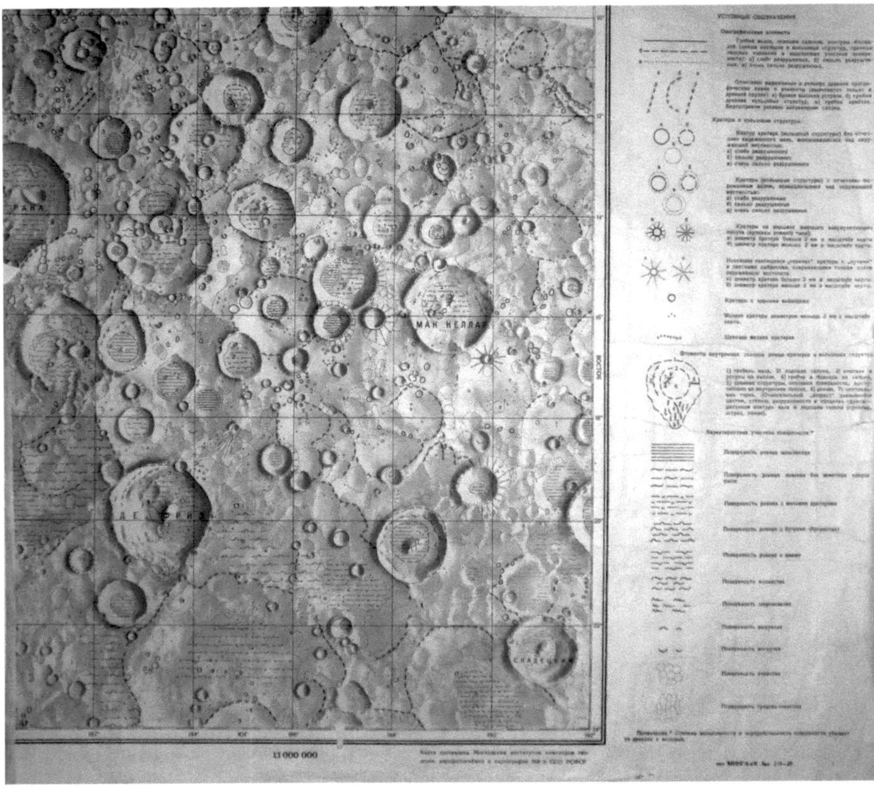

Fig. 2.19 Map of the Moon: part of the equatorial region of the farside (MIIGAiK)

surface: the half-tone colour wash (by A. S. Tolstoukhov 1978) and the same combined with lines and contours (Volkov 1978). An example sheet is shown in Fig. 2.19.

On the basis of the half-tone map from MIIGAiK (editors: B. Krasnopevtseva and K. Shingareva), a multilingual map at 1:12,800,000 scale was compiled in Hungary (Map of the Moon, Multilingual edition, 2002), edited by Henrik Hargitai as part of a series of multilingual maps (Fig. 2.20). The reverse side of the map gives data on the structure of the lunar surface in five languages: English, Czech, Polish, Croatian and Bulgarian. The map is saturated with information, especially where features are named in all five languages. The map was proposed for publication in Budapest, Moscow, Prague, Warsaw, Krakow, Zagreb and Sofia.

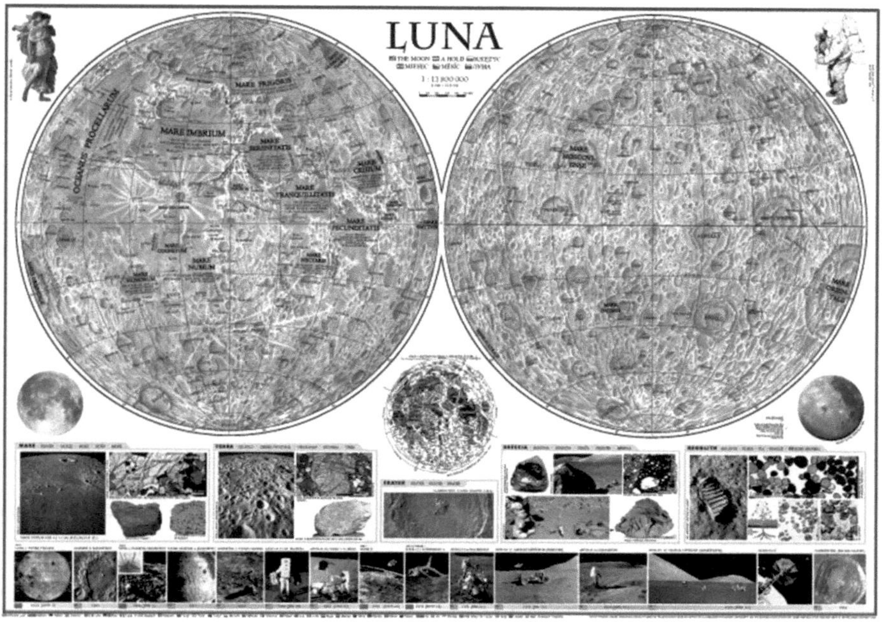

Fig. 2.20 Multilingual map of the Moon (2002)

References

(1977). *Karta Obratnoy Storonu Lunu na rayon s'emok s avtomaticheskoy stantsii Zond 8. 1 :1 000 000*. Moskva: Izdatelstvo MIIGAiK.

Arthur, D. W. G., Agnieray, A. P., Pellicory, R. H., Wood, C. A., Weller, T. et al. (1963). Communication Lunar and Planetary Laboratory Vol. 2, No. 30.

Arthur, D. W. G., Agnieray, A. P., Pellicory, R. H., Wood, C. A., Weller, T. et al. (1964). Communication Lunar and Planetary Laboratory Vol. 3, No. 40.

Arthur, D. W. G., Agnieray, A. P., Pellicory, R. H., Wood, C. A., Weller, T. et al. (1965). Communication Lunar and Planetary Laboratory Vol. 3, No. 50.

Arthur, D. W. G., Agnieray, A. P., Pellicory, R. H., Wood, C. A., Weller, T. et al. (1966). Communication Lunar and Planetary Laboratory Vol. 5, part 1, No. 70.

Atlas Obratnoy Storonu Lunu. (1960). *Nauchnue rukovoditeli Barabashov, N.P., Mikhailov, A.A., Lipskiy, Yu. N.* Moskva: Izdatel'stvo Akademii Nauk SSSR.

Atlas Obratnoy Storonu Lunu. Chast' 2. (1967). *Nauchnuiy rukovoditel Lipskiy, Yu. N.* Moskva: Izdatel'stvo Akademii Nauk SSSR.

Atlas Obratnoy Storonu Lunu. Chast' 3. (1975). *Nauchnuiy rukovoditel Lipskiy, Yu. N.* Moskva: Nauka.

Boginskiy, V. M., Rodionova, Zh. F., Myshetskaya, E. N. (1990). Oput Sozdaniya Globusa Lunu na Termoplastichnom materiale. *Geodesiya I Kartografiya, 6*, 28–34.

Brian Harvey. (2007, June 2). *Mapping the Moon*. Paper presented to the British Interplanetary Society, BIS 2007.

Globus Lunu 1:10 000 000. (1967). GAISH, TGS SSSR, Nauchnuiy rukovoditel Lipskiy, Yu.N. Izdatel'stvo "Nauka". Moskva.

Globus Lunu 1:10 000 000. (1969). GAISH, TGS SSSR, Nauchnuiy rukovoditel Lipskiy, Yu.N. Izdatel'stvo "Nauka". Moskva.

Globus Lunu 1:10 000 000. (1974). GAISH, TGS SSSR, Nauchnuiy rukovoditel Lipskiy, Yu.N. Izdatel'stvo "Nauka". Moskva.
Globus Lunu 1:10 000 000. (1979). GAISH, TGS SSSR, Nauchnuiy rukovoditel Lipskiy, Yu.N. Izdatel'stvo "Nauka". Moskva.
Karta Lunu 1:10 000 000. (1985). GAISH, TGS SSSR. "Nauka". Moskva.
Lipskiy, Y. N., & Chikmachev, V. I. (1969). Kartograficheskaya pivyazka snimkov AMS Zond 3 sposobom rektifikatsii po orientiruyuschim tochkam. *Astronomicheskiy Zhurnal, 46*(6), 643–651.
Lipskiy, Y. N., Nikonov, V. A., Rodionova, J. F., Chikmachev, V. I., Shevchenko, V. V., Lavrova, N. B., et al. (1977). Materialu Sostavleniya Polnoiy Kartu Lunu 1:5 000 000 3-ego Izdaniya. *Soobscheniya GAISH, 204*. 60 p.
Lipskiy, J. N., Nikonov, V. A., & Skobeleva, T. P. (1973). *Edinaya Sistema Selenodezicheskih Koordinat iz Devyati Katalogov na Vidimom polusharii Lunu "Nauka"*. Moskva.
Lipskiy, Yu. N., Pskovskiy, Yu. P, Gurshteyn, A. A., Shevchenko, V. V., & Pospergelis, M. M. (1966). Tekuschie problemu morfologii poverhnosti Lunu. *Kosmicheskie Issledovaniya, 4*, Vupusk 6, 912–922.
Lipskiy, Yu. N., & Rodionova, Zh. F. (1978). *Kartografirovanie Lunu. V kn. Uspehi Sovetskogo Soyuza v issledovanii kosmicheskogo prostranstva*. Vtoroye kosmicheskoye desyatiletie 1967–1977 (pp. 406–434).
Lisina, L. R., & Shevchenko, V. V. (1965). Opredeleniye Selenograficheskih Koordinat Tochek Lunnoiy Poverhnosti I Rasstoyaniy mezhdu nimi pri Fotometricheskom Sposobe issledovahiya uglov naklona. *Sbornik Figura i Dvizhenie Lunu* (pp. 80–91). Kiev: Naukova Dumka.
Polnaya Karta Lunu 1:5 000 000. (1967). GAISH, TGS SSSR. Nauchnuiy rukovoditel Lipskiy, Yu. N. Izdatel'stvo "Nauka". Moskva.
Polnaya Karta Lunu 1:5 000 000. (1969). GAISH, TGS SSSR. Nauchnuiy rukovoditel Lipskiy, Yu. N. Izdatel'stvo "Nauka". Moskva.
Polnaya Karta Lunu 1:5 000 000. (1979). GAISH, TGS SSSR. Nauchnuiy rukovoditel Lipskiy, Yu. N. Izdatel'stvo "Nauka". Moskva.
Polnaya Karta Lunu 1:10 000 000. (1979). Nauchnuiy rukovoditel Lipskiy, Yu.N. Izdatel'stvo GAISH, TGS SSSR. "Nauka". Moskva.
Rodionova, Zh. F. (1969). Izmerenie Ploschadey Lunnuh Morey i Talassoidov. *Astronomicheskiy tsirkulyar AN SSSR, 538*, 1–3.
Rodionova, Zh. F. (1972). Gipsometricheskiye Osobennosti Vidimogo Polushariya Lunu. *Astronomicheskiy Zhurnal, 49*, Vupusk 3, 121–127.
Rodionova, Zh. F. (1983). Obzornoye Kartografirovaniye Lunu. Izvestiya Vuzov. *Geodesiya i Aerofotosyiomka, 1*, 24–33.
Rodionova, Zh. F., Karlov, A. A., Skobeleva, T. P., Konotopskaya, E. V., Shevchenko, V. V., Kozubskiy, K. E., et al. (1987). *Morfologicheskiy Katalog Kraterov Lunu* (173 s). Moskva: Izdatelstvo Moskovskogo Universiteta.
Shevchenko, V. V. (1963a). Zvezdnaya Karta Blizpolusnoy Zonu Seernogo Polusharia Lunu. *Izvestiya Vuzov. Geodesiya i Aerofotosyiomka, 1*, 87–91.
Shevchenko, V. V. (1963b). Vidimue Peremescheniya Zemli na nebe Lunu. *Kosmicheskie Issledovaniya, I*, Vupusk 2, 216–220.
Shevchenko, V. V. (1968). Ispol'zovanie Zemli dlya selenograficheskoy privyazki lunnuh izobrazheniy, poluchennuh iz kosmosa. Kosmicheskie issledovaniya. T.6, N.6 (924).
Shevchenko, V. V. (1977). *Soderzhaniye i Stratigrafiya Lunnuh porod po Dannum ob ih Albedo*. Sovetsko-Amerikanskaya Konferentsiya po Kosmohimii Lulu i Planet (pp. 253–258).
Shevchenko, V. V. (1980). *Sovremennaya Selenografiya*. Moskva: Nauka.
Shingareva, K. B., & Burba, G. A. (1977). *Lunnaya nomenklatura. Obratnaya Storona Lunu 1961–1973*. Moskva: Nauka.
Tolstoukhov, A. S. (1978). *Nekotorue Osobennosti Otobrazheniya Lunu Metodom Polutonovoy Otmuvki* (pp. 63–66).
Volkov, N. M. (1978). *Tehnologiya Kartografirovaniya Lunu*. V kn. Kartografirovanie Lulu i Marsa (pp. 58–63).
Westfall, J. E. (1970). The areas of the lunar lowlands. *Strolling Astronomer, 22*(7–8).

Chapter 3
Cartography of the Lunar Nearside

Abstract The Soviet spacecraft *Luna 9* made the first ever soft landing on the lunar surface at the western edge of Oceanus Procellarum. The soft landing was elaborated over the course of several spacecraft launches. The soft landing of *Luna 9* and its transmission of unique panoramic images of the lunar microrelief opened a new epoch in the study of the cosmos. The first topographic scheme for parts of the lunar surface using the data from *Luna 9* was compiled at MIIGAiK in 1966. The schematic map of the landing area was constructed at 1:40 scale. In 1966, the *Map of the Moon: equatorial zone of the nearside* was compiled at GAISh and TsNIIGAiK in Mercator cylindrical projection at 1:1,000,000 scale for the region ±8° in latitude and ±70° in longitude. A photomap of the lunar nearside at 1:5,000,000 scale was compiled in 1967. Taking this map as a basis, in 2014 we produced a modern version at 1:8,000,000 scale, marking all the landing sites for both unmanned and manned spacecrafts. *Lunokhod 1* was the first ever planetary rover. It operated on the surface of the Moon for 10.5 months (11 lunar days). A topographic plan, originally compiled at 1:200 scale, of a part of the route has the system of symbols developed in IKI. The DEM and orthoimage were used to map the *Lunokhod 1*'s complete traverse in 2012 in MIIGAiK. The IAU approved the naming of small craters along the route taken by *Lunokhod 1* with the first names of members of the team controlling the lunokhod.

First Topographic Map Scheme of a Region of the Lunar Surface

On February 3rd 1966, the Soviet spacecraft *Luna 9* made the first ever soft landing on the lunar surface at the western edge of Oceanus Procellarum. This successful experiment was the outcome of the vast effort of designers, engineers and workers, whose hands created the spacecraft. In December 1961, then unknown chief designer Sergei Pavlovich Korolev gathered the famous astronomers A. G. Masevich, D. Y. Martynov, Yu. N. Lipsky, M. M. Kobrin, A. A. Mikhailov, N. P. Barabashov and V. S. Troitskii in the Academy of Sciences to find out whether the surface of our natural satellite, the Moon, was hard, or whether it was covered in a thick layer of dust into which a spacecraft could sink.

V. Shevchenko et al., *Lunar and Planetary Cartography in Russia*, Astrophysics and Space Science Library 425, DOI 10.1007/978-3-319-21039-1_3

The first information about the structure and physical properties of the upper surface of the Moon were obtained by astronomers with Earth-based measurements of its optical, thermal and radio emission. Later in situ measurements showed that the findings were rather close to the truth. The leading role in the research of the lunar surface properties was played by the Soviet astronomers N. P. Barabashov, A. V. Markov, V. S. Troitskii, V. G. Fesenkov, Yu. N. Lipsky, V. V. Sharonov and N. N. Sytinskaya, who had begun their work long before the first space flights.

Interpretation and simulation of the derived optical characteristics of the lunar surface led to the fundamental conclusion that the outer layer of the cover of the Moon had a high porosity. Study of the infrared radiation from the lunar surface material showed that its thermal conductivity was hundreds of times lower than that of Earth rocks, confirming the idea of its high porosity. An evaluation of the properties of the lunar covering at greater depths was first done by V. S. Troitskii (Troitskiy 1963) and his colleagues at the Radiophysics Institute in Gorkii (known today as Nizhnii Novgorod). By monitoring the radio emission from the Moon, they established that the mean density of the lunar covering gradually increases. Thus, at 4 cm depth, it can be 0.6 g/cm^3; at 3 m depth, be 1 g/cm^3 and at 6 m depth, 1.5–2 g/cm^3. These observations showed that the upper layer of the Moon was a hard material, the density of which was about half that of water. It likewise became clear that the lunar surface conducted heat about 50 times better than a layer of fine dust, which completely negated the supposition of a thick layer of dust. Putting this evidence together, the astronomers concluded that the lunar surface resembles the familiar pumice stone.

This finding was in full agreement with the so-called meteor-slag theory of the structure of the surface layer of the Moon, formulated by Sharonov and Sytinskaya back in 1959. At the same time, J. Kuiper conducted his own research for NASA and came up with a load-bearing capacity for the upper layers of the lunar soil of 1 kg/cm^2. Thus, arose the much repeated journalistic tale that Korolev personally decreed that the Moon should be considered solid.

The decision was taken at a meeting, where the results from Troitskii (NIRFI), Barabashev (Kharkov University Observatory) and Sharonov and Sutinskaya (Leningrad University) were considered (Barabashov 1963; Sharonov 1963; Sutinskaya 1963). On the other hand, the hypothesis of a many-kilometre covering of fine dust, supported by T. Gold on the basis of speculative theoretical constructions, was not supported by the majority of astronomers, and the idea's strongest critic was G. Kuiper. According to this hypothesis, the UV and corpuscular radiation from the Sun destroy the crystal lattice of minerals, meteor strikes grind and mix the lunar soil, whereby a layer of dust is produced which could grow to depths of several kilometres. This continuous dust layer should not remain where it is formed, but migrate from elevated areas to lowlands. The model became widely known not through professionals, but from a science fiction novel by Arthur C. Clarke.

The soft landing was elaborated over the course of several spacecraft launches. At the beginning of April 1963, *Luna 4* was launched successfully, first going into orbit about Earth and then, having overcome the Earth's gravity, was to go into orbit around the Moon. It passed the Moon, however, at a distance of 8500 km. *Luna 5*, launched on 9th May 1965, crashed on the surface of the Moon to the south-west of

Copernicus Crater. *Luna 6* passed over the surface of the Moon at a distance of 1627 km in June 1965. In that year, the only successful flight was *Zond 3*, which photographed part of the lunar farside. As described in Chap. 2, 25 photographs of the lunar farside from both this mission and *Luna 3* served as the basis for the first complete map of the Moon. At the beginning of October 1965, *Luna 8* suffered a failure. In autonomous mode, it successfully completed a braking manoeuvre on approaching the lunar surface, but when its protective airbag began to inflate, it was punctured by a small plastic arm, and the spacecraft broke on impact. The idea of making a soft landing on the Moon using an "inflatable ball", inside which all the apparatus would be hidden was Korolev's. Chertok later said, "When I looked at Sergei Pavlovich, I didn't know what to pity more: the lunar spacecraft lying in pieces in the dust somewhere out there, four hundred thousand kilometres away, or this great, strong man, who was disappointed beyond measure" (Chertok 1966).

When it came to reporting the failure of *Luna 8* at the Kremlin, Korolev did it himself, although formally it should have been Chertok. Chertok believed that the Chief Designer saved the soft landing project with his presentation, which might otherwise have been cancelled. Korolev said, "Keep in mind: this is a learning process. Yes, we make mistakes—even stupid ones leading to serious failures. But from launch to launch, we are approaching success. We are so close to victory, that I can guarantee the next attempt will be successful."

Soon after, all the work on the creation of interplanetary spacecraft was transferred from OKB-1, headed by Korolev, to the Lavochkin engineering plant, and Georgii Nikolaevich Babakin was nominated Chief Designer. Probes which landed on the Moon and returned lunar samples to Earth, which flew as the first artificial satellites of the Moon, and which were sent to study other planets were developed under his leadership. Today, there are craters named after Babakin on both the Moon and Mars.

The first spacecraft made under Babakin was *Luna 9* (Fig. 3.1). *Luna 9* made a soft landing at the edge of Oceanus Procellarum (at 7°8′N 64°32′W) between craters Cavalerius and Galilaei. The soft landing was made 20 days after the fateful operation which unexpectedly took the life of Sergei Pavlovich. His method of soft landing, demonstrated by *Luna 9*, was later adopted by American designers and for landing spacecraft on Mars. By a decision of the International Astronomical Union, the region around the landing site of *Luna 9* received the special name Planitia Descensus (www.planetarynames.wr.usgs.gov).

Inside the body of the probe, which consisted of two hemispheres, was a frame with the radio communication apparatus, a chemical battery, electronic program and timing devices and scientific instruments. A television system provided a direct transmission to Earth of a panoramic view of the lunar landscape (Fig. 3.2). The petal and rod antennae and the mirrors were in a folded configuration on landing. After touchdown, the program-timer triggered the release of pyrotechnic bolts, opening up the antennae. The probe weighed 100 kg, with the antennae being 112 cm. The diameter of the probe after opening was 160 cm (Fig. 3.1). The transmission of a single panorama lasted 100 minutes: there were seven communication sessions in all, with a total duration of more than 8 hours (Pervuye panoramu lunnoy poverhnosti 1966).

Fig. 3.1 Model of *Luna 9* at the Lavochkin Museum (Photo: Zh. F. Rodionova)

Fig. 3.2 Part of the panorama of the lunar surface seen from the *Luna 9* landing site

The soft landing of the Soviet *Luna 9* and its transmission of unique panoramic images of the lunar microrelief opened a new epoch in the study of the cosmos. The experiment revealed the microstructure of the lunar soil, discovered the presence of rocks at the surface and established that the Moon was not covered by a significant layer of dust (Fig. 3.2).

Three complete panoramas of the vicinity were received with the Sun at 7°, 14° and 27° above the horizon, and a partial fourth panorama with the Sun at 42°. By comparing the images with differing illumination, it was possible to assess the microrelief and structure of the lunar soil from the changing shadow lengths (Pervuye panoramu lunnoy poverhnosti 1969). Additional data on the properties of the lunar soil were obtained by *Luna 13* which had special sensors mounted on a 1.5 m arm for measuring the mechanical properties of the soil: a stamping device and a density sensor using radiation.

The first topographic scheme for parts of the lunar surface using the data from *Luna 9* was compiled at MIIGAiK in 1966. The schematic map of the landing area was constructed at 1:40 scale (Fig. 3.3; Luna 13 – Fig. 3.4). Contours marked the heights in centimetres with 5 cm intervals, and craters and rocks were numbered. The heights ranged from −2 to −59 cm. The radial distance from the probe was marked at 1 m intervals, with azimuthal lines every 20° from the centre of the panoramas. 86 crater and 74 rocks were identified. Schematic map of the landing area of Luna 13 is shown in Fig. 3.4

Map of the Moon: Equatorial Zone of the Nearside

In 1966, the *Map of the Moon: equatorial zone of the nearside* was compiled at GAISh and TsNIIGAiK in Mercator cylindrical projection at 1:1,000,000 scale for the region ±8° in latitude and ±70° in longitude (Fig. 3.5). The map was issued in 1968 (Karta Luny. Ekvatorial'naya zona vidimogo polushariya 1968). In direct cylindrical projections, there is no distortion of angles so that features on the map retain their true shape.

A series of 44 map sheets of the nearside at 1:1,000,000 scale was published in the USA over the years 1960–1968. The exterior layout of our maps was kept the same as those published in the USA, with the intent that such a design could become an international standard for maps at this scale. The relief of the lunar surface was illustrated with a half-tone shading (Rodionova 1969). The solar incidence angle varied in the illustration to better express both details of relief that appear at low incidence angles (gentle ridges in the maria, domes, fissures and grooves) and those which appear at high angles (ray systems, mare-type features). Different colour shades define regions differing in albedo (maria and highland regions). Within the maria there are many "islands" which are brighter or darker than the vicinity: these are expressed on the map (Fig. 3.6).

The colour range for the maps was chosen with the aim of achieving the greatest expression of relief. The colour wash was carried out by the cartographers P. K. Kolaevii and A. C. Drago of TsNIIGAiK. This map was the first to show the location of the *Luna 9*'s soft landing, named Planitia Descensus, confirmed by the IAU General Assembly in 1970. In the 1968 edition of the map, the scale is true at the equator. The Goloseevo Observatory control point network of 500 points and a mean precision of 8′ in longitude and 3′ in latitude were used for positioning.

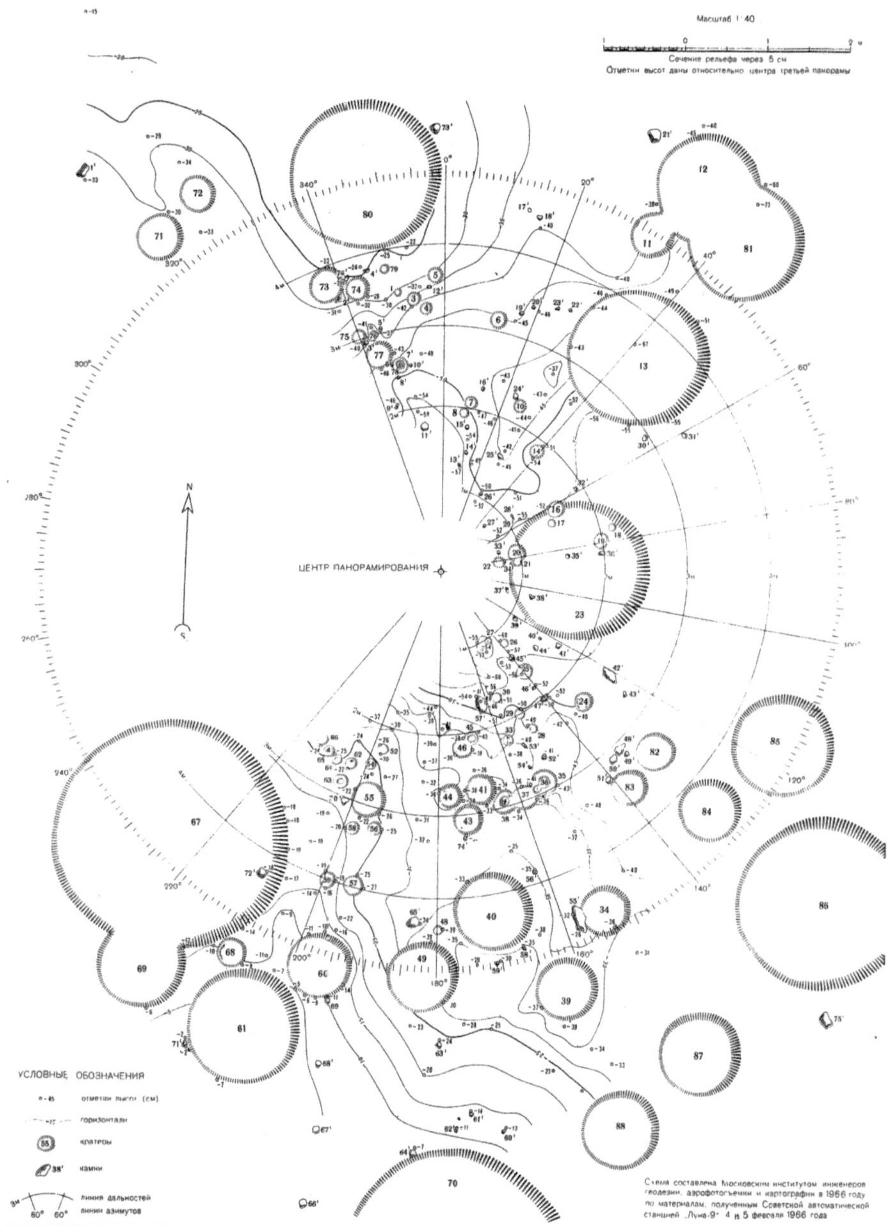

Fig. 3.3 Schematic map of the landing area of Luna 9

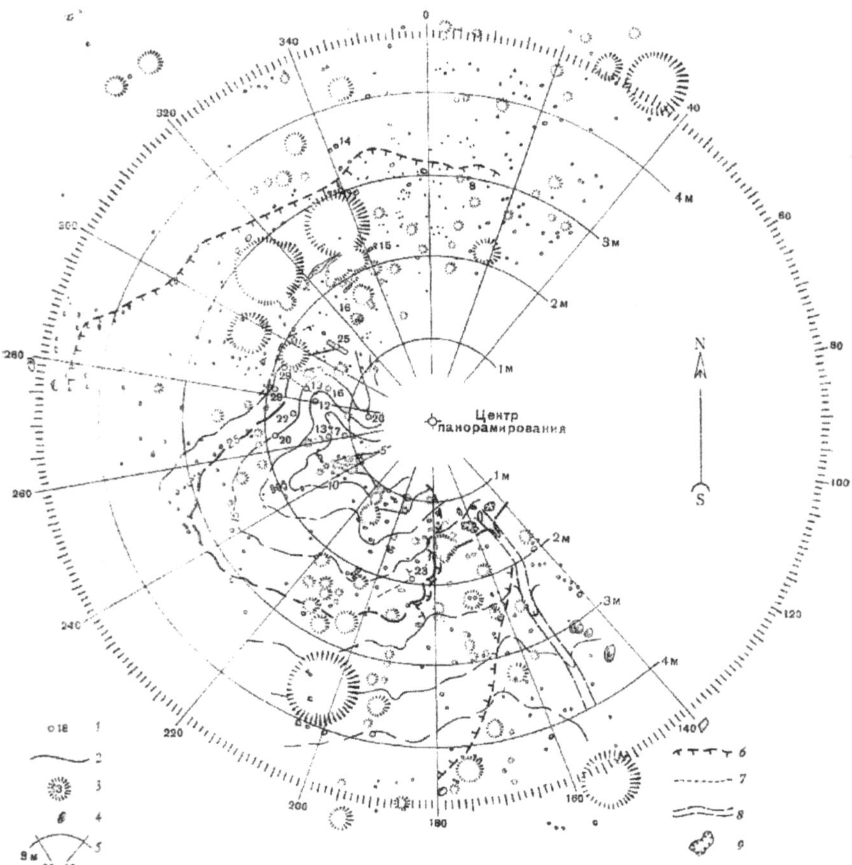

Fig. 3.4 Schematic map of the landing area of Luna 13

Photomap of the Nearside Lunar Hemisphere

A photomap of the lunar nearside at 1:5,000,000 scale (Fig. 3.7) was compiled at GAISh and TGS USSR in 1967 on the basis of original photographs obtained by Soviet and foreign observatories and material from photographic atlases of the Moon (Fotokarta vidimogo polushariya Luny 1967). The map was constructed in oblique positive external projection for positive values of libration in latitude and longitude close to the maximum, allowing Mare Humboldtianum, Mare Marginis and Mare Smythii to be shown. The difference between external perspective projection and positive projection is that the plane of the map is projected onto the part of the sphere which faces the projection point (Volkov 1964; Bugayevskiy 1998). Photographs of the Moon are obtained with positive projection, whether taken from space or from the Earth.

Fig. 3.5 Scheme of sheets for *Map of the Moon: equatorial zone of the nearside* (1968)

For the east part of the map, telescopic images were used which had been obtained close to the first quarter Moon phase, and the west part was made from images made close to the last quarter phase. This enabled an even clarity of relief, but the direction of shadows from mountains and craters are different. The central point of the photomap is at 6.1°N 5.3°E. This allows a good view of the northern polar area and the east limb. The south pole is not visible with these values.

Taking this map as a basis, in 2014 we produced a modern version at 1:8,000,000 scale (Fig. 3.8), marking all the landing sites for both unmanned and manned spacecraft (Agamalyan et al. 2014). The map includes a table with a short description of the scientific results obtained by the spacecraft, beginning from *Luna 2*, launched by the USSR in 1959, up until the Chinese spacecraft *Chang'e 3* which placed the rover *Yutu* on Mare Imbrium in 2013. In total, there have been 23 lunar landings. Four made hard landings: *Luna 2* and *Ranger 7, 8, 9*. The first soft landing was *Luna 9*, and the first American soft landing was *Surveyor 1*. Six lunar modules brought 12 American astronauts to the surface, and returned 379 kg of lunar samples. The unmanned soviet spacecraft *Luna 16, 20, 24* brought back lunar soil, and *Luna 17, 21* delivered *lunokhod* rovers to the Moon. All these locations are marked on the map.

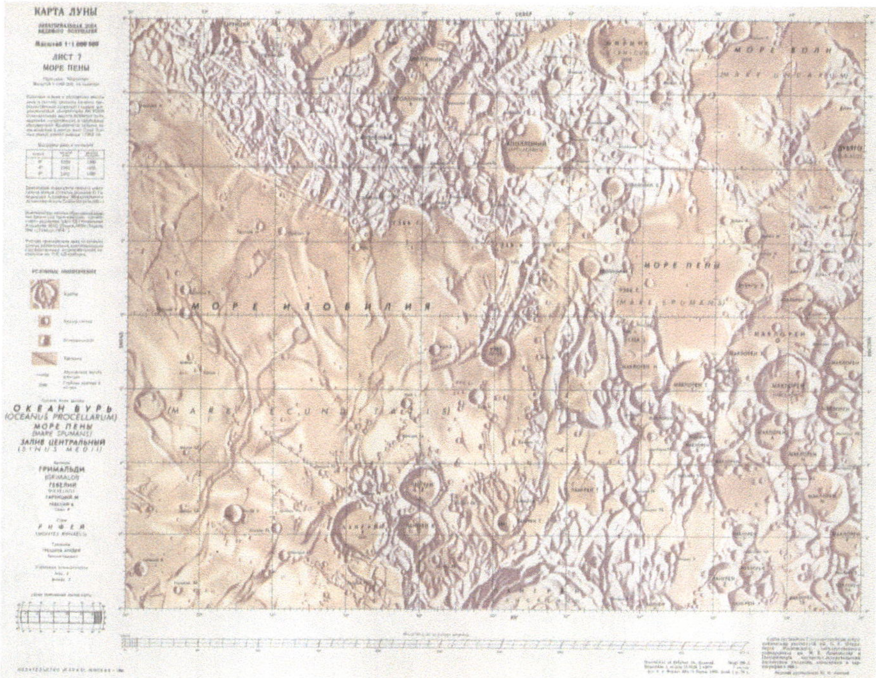

Fig. 3.6 Sheet 7 of *Map of the Moon* at 1:1,000,000 scale (1968)

The map includes new names and physical characteristics of the Moon. The diameter of the lunar disk on the map is 483 mm. Russian feature names are marked according to the catalogue maintained at GAISh: on the map, only those of larger features are marked. Among the new names approved by the IAU are craters Keldysh, Glushko and Yangel'. Many smaller craters on the Moon are denoted with first names used in different countries: this map includes only a few of these, e.g., Natasha Crater.

Map of the Moon: The Nearside in Telescopic View

In 1967, the *Map of the Moon: the nearside in telescopic view* at 1:5,000,000 scale (edited by V. A. Bronshten was published for amateur astronomers (Fig. 3.9). The map was constructed by engineer and cartographer I. I. Katyaev in transverse (equatorial) orthographic projection from the photographic atlas of the Moon by Kuiper (1960) and the catalogue of lunar features IAU (Blagg and Muller 1935).

The Moon is displayed inverted, as seen in the telescope, for zero libration in latitude and longitude. Mare and highland regions, mountain ranges and craters, are marked with different colours. Conventional signs indicates the landing sites of the

Fig. 3.7 Photomap of the nearside of the Moon (1967)

soviet spacecraft *Luna 2, 5, 7, 8, 9, 13* and the American spacecraft *Ranger 6, 7, 8, 9* and *Surveyor 1*. The map booklet, prepared by V. A. Shishakov, gives a detailed description of the motion of the Moon, the conditions of visibility for changing Moon phase and explains in detail which details can be observed in a particular phase (Karta Luny. Vidimaya storona v teleskopicheskom izobrazhenii 1967).

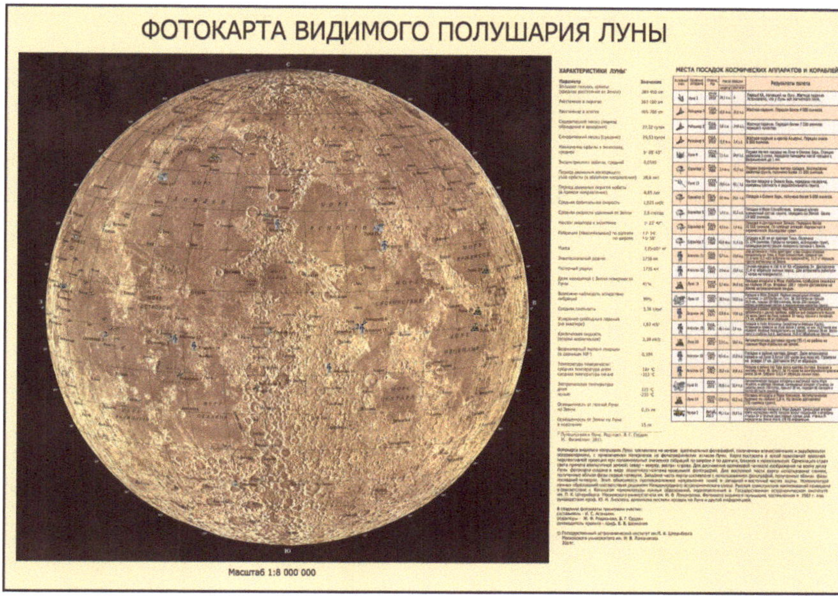

Fig. 3.8 Photomap of the nearside hemisphere of the Moon (2014)

The Lunokhods

Lunokhod 1 was the first ever planetary rover. It operated on the surface of the Moon for 10.5 months (11 lunar days). It was delivered to the surface of the Moon on 17th November 1970 by the soviet spacecraft *Luna 17* and was working until 4th October 1971 (Fig. 3.11). It was designed to study the characteristics of the lunar surface, the radiation and x-ray environment and the chemical composition and properties of the lunar soil. The apparatus was created in the design office of the Lavochkin Engineering Plant directed by Babakin, and the chassis was built at VNIITransMash directed by A. L. Kemurdzhian. *Lunokhod 1* travelled 10,540 m, and transmitted 211 lunar panoramas and 25,000 images. A part of one of the panoramas is shown in Fig. 3.10. The physical and mechanical properties of the surface layer of soil were tested at 500 points along the route. A chemical analysis was carried out at 25 locations.

The main difficulty in controlling the lunokhod was the time delay in the radio communication: it took 2 seconds for the signal to get to the Moon and back, and the small-frame television camera could send images at rates of one every 4–20 seconds. Consequently, there was a combined delay of up to 24 seconds. Details of the operation of the rover in lunar conditions, the study of the topography of the region and the chemical composition of the soil, as well as geomorphological research were presented in two volumes of *Lunokhod 1: Mobile Laboratory on the Moon*, published in 1977 and 1978 (Peredvizhnaya Laboratoriya na Lune. Lunokhod 1. Basrukov ed., 1971, 1978).

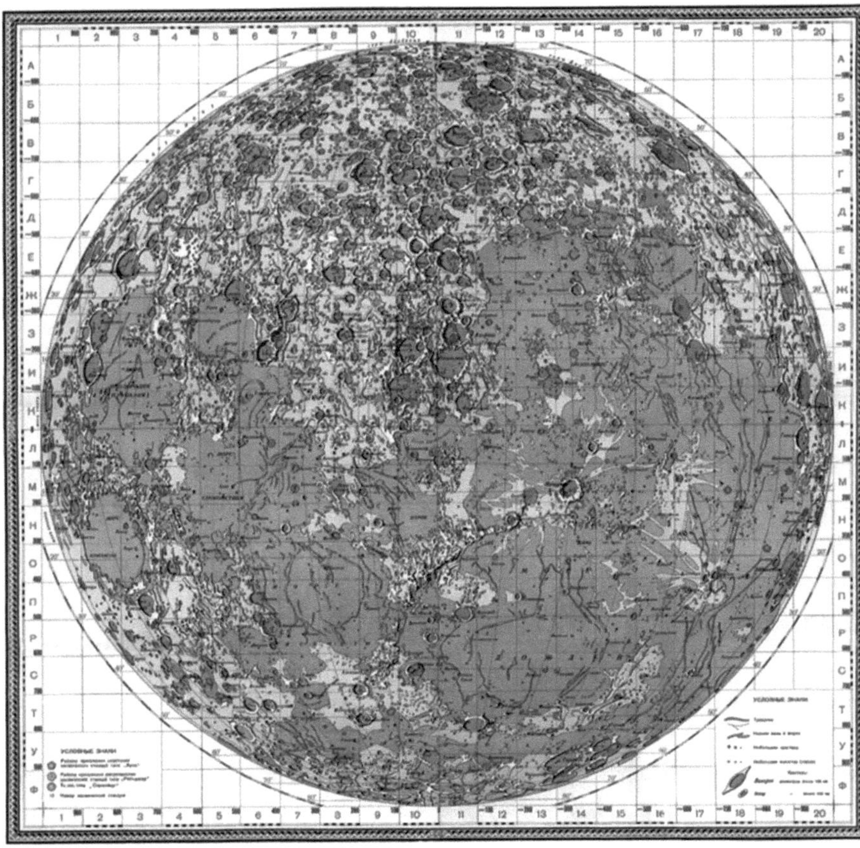

Fig. 3.9 Map of the Moon: the nearside in telescopic view (1967)

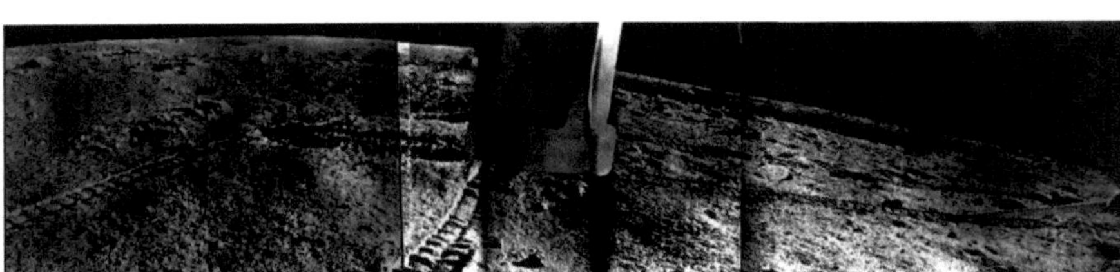

Fig. 3.10 Part of a panorama of the lunar surface transmitted by *Lunokhod 1*

In March 2010, Albert Abdrakhimov rediscovered *Lunokhod 1* in images from *Lunar Reconnaissance Orbiter* (*LRO*) (Abdrakhimov et al. 2011). The *Luna 17* landing platform as well as the roving vehicles at their final resting positions could be clearly identified, and the rover tracks are discernable in most areas. Figure 3.11 shows the place of site soviet Luna 17 (Lunohod 1).

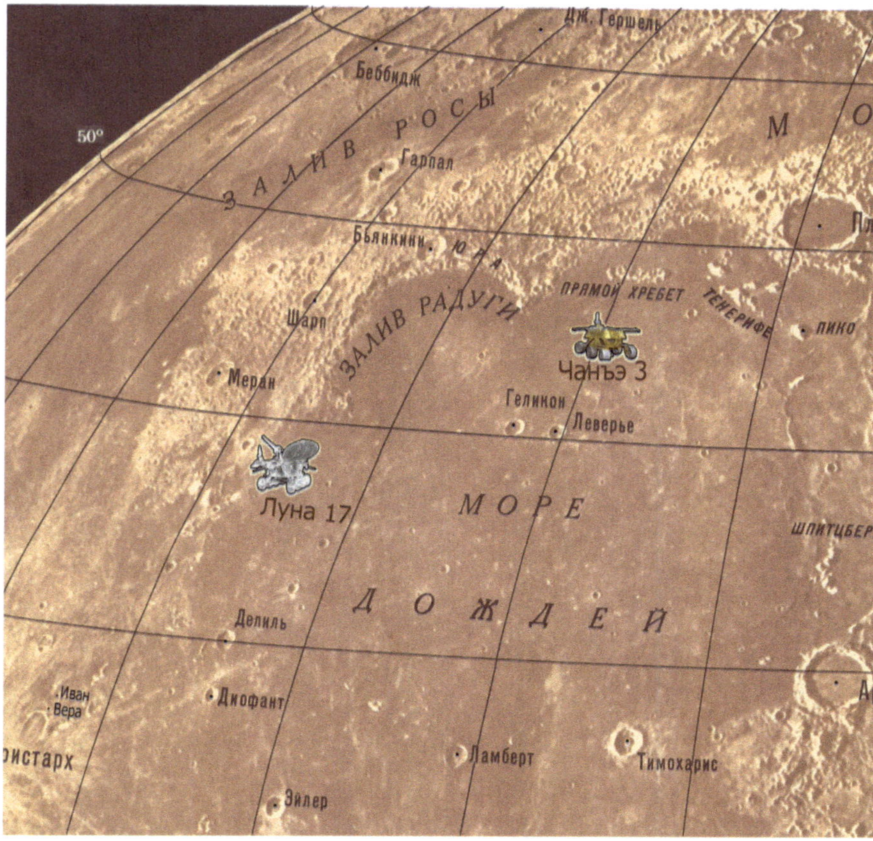

Fig. 3.11 Fragment of photomap of the nearside hemisphere of the Moon (2014) with the places of site soviet Luna 17 (Lunokhod 1) and Chinese Chang'e 3

A topographic plan, originally compiled at 1:200 scale, of a part of the route is given in Fig. 3.12. The system of symbols developed for the topographic plans (Shingareva and Burba 1978) allowed them to be expanded or reduced in scale according to need. For example, if it was required to reflect the density of craters, mounds, rocks or debris, the number of symbols per unit area could be varied. Rocks with a particular characteristic could have a different hatching or fill. Additional gradations of steepness could be represented by various combination of lines and dots in the drawn lines and likewise differing line thicknesses for the crater contours.

Lunokhod 1 was equipped with a corner reflector for precise measurements of the distance to the Moon. During the first 1.5 years on the lunar surface, about 20 successful measurements were made using the reflector, but then its precise position was lost. On 22nd April 2010, a group of US researchers led by Tom Murphy reported that they had managed to detect a return signal from the reflector

Fig. 3.12 Topographic plan of a part of the *Lunokhod 1* route

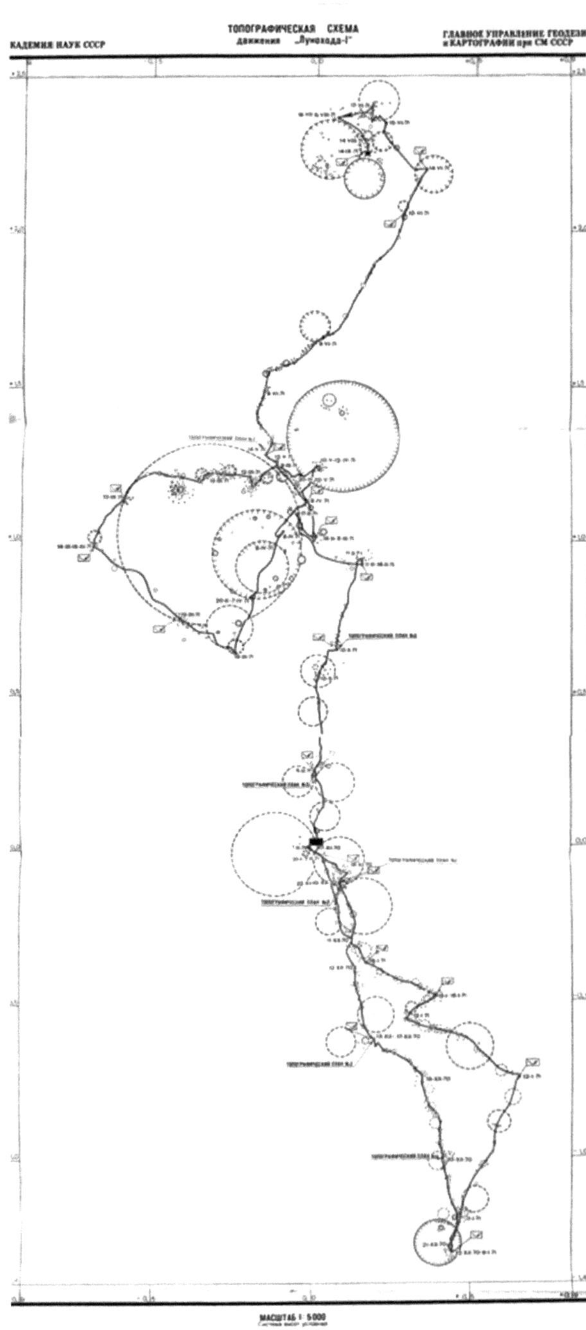

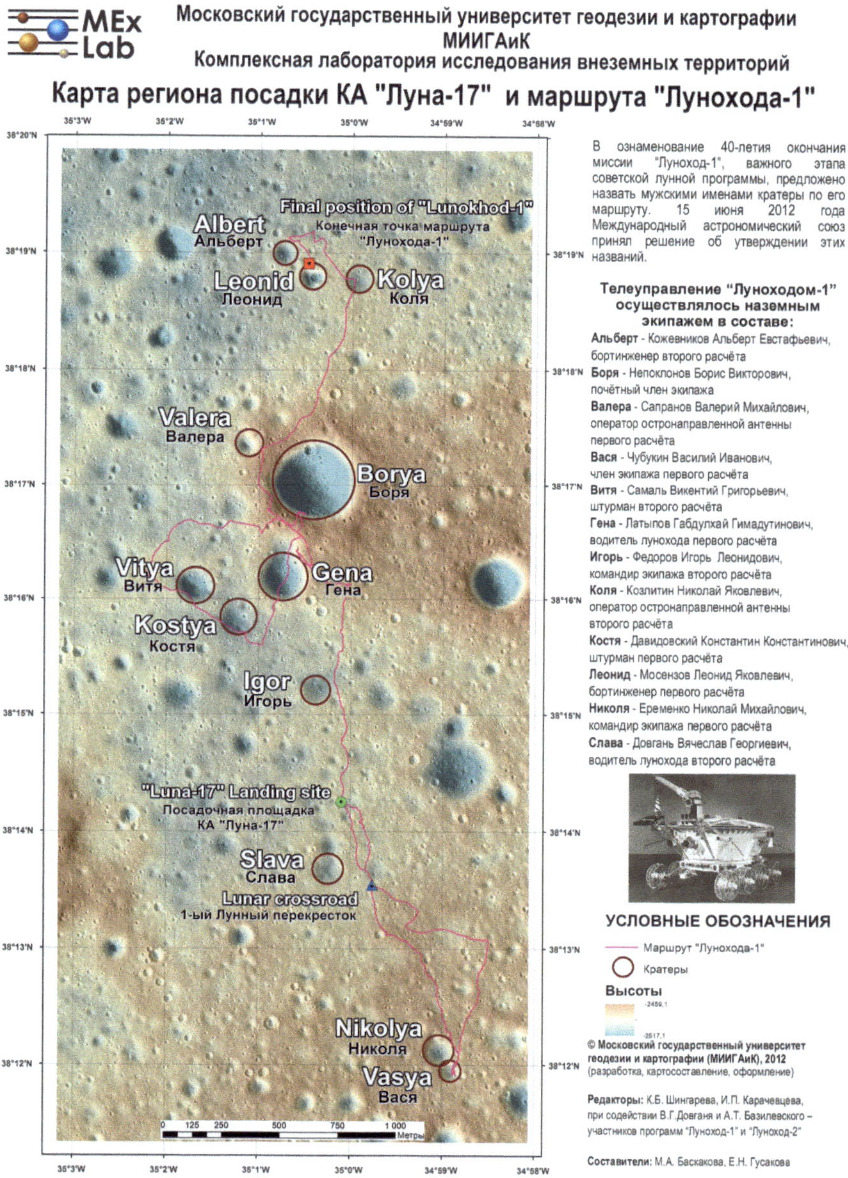

Fig. 3.13 Map of the landing site of *Luna 17* and the route taken by *Lunokhod 1* compiled at MIIGAiK. The craters along the route were given the first names of members of the team controlling the lunokhod

illuminated by a laser pulse. *Lunokhod 1*'s final resting position was determined to be 38.32°N 35.01°W.

High-resolution DEMs and orthoimages were produced by photogrammetric processing of *Lunar Reconnaissance Orbiter Camera Narrow Angle Camera* (*LROC NAC*) stereo images with spatial resolution of 0.5 m/pixel (M150749234, M150756018). A topographic profile along the traverse showed that the surface varied in height from −2488.4 to −2462.3 m (Gusakova et al. 2012, 2013). The coordinates of the surveying points from where stereo images had been acquired by *Lunokhod 1* were determined. A catalogue of about 45,000 craters with diameters and depths was obtained from the DEM. The DEM and orthoimage were used to map the *Lunokhod 1*'s complete traverse (Karachevtseva et al. 2010, 2011, 2013). In 2012, the IAU approved the naming of small craters along the route taken by *Lunokhod 1* with the first names of members of the team controlling the lunokhod (Fig. 3.13).

The exact location of *Lunokhod 2* (route taken by Lunohod 2 is shown in Fig. 3.14) was soon found in a similar fashion in *LRO* images by V. Kaidash and C. Gerasimenko of Kharkov University Observatory (25.83°N 30.91°W) (Fig. 3.14). In December 1993, NPO Lavochkin had auctioned off *Lunokhod 2* together with *Luna 21* at Sotheby's in New York for $68,000 to the son of astronaut-businessman Richard Harriot, who made a flight to the International Space Station as a tourist on a Soyuz TMA-13.

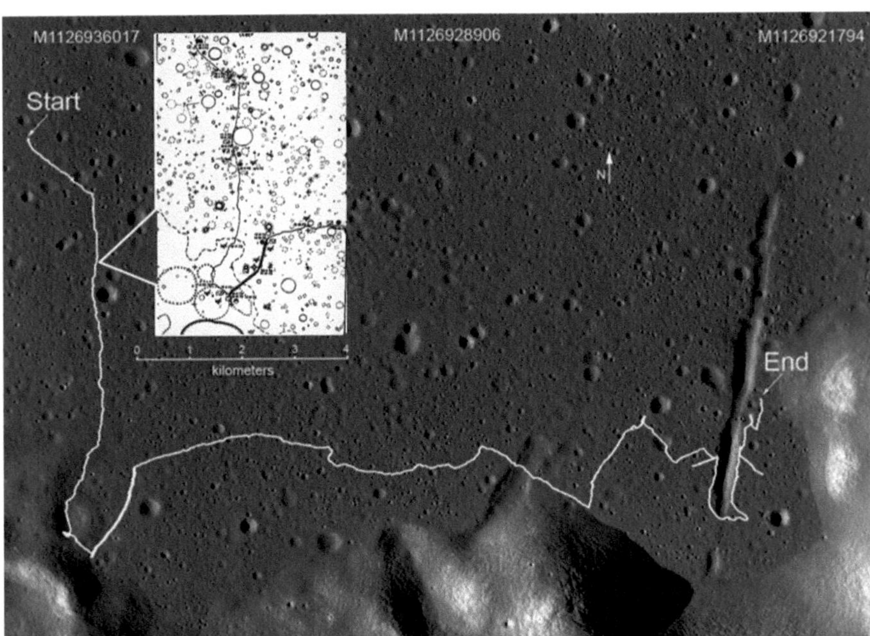

Fig. 3.14 Route taken by *Lunokhod 2* on a mosaic of six *LRO* images and a part of topographic plan (Rodionov et al. 1973)

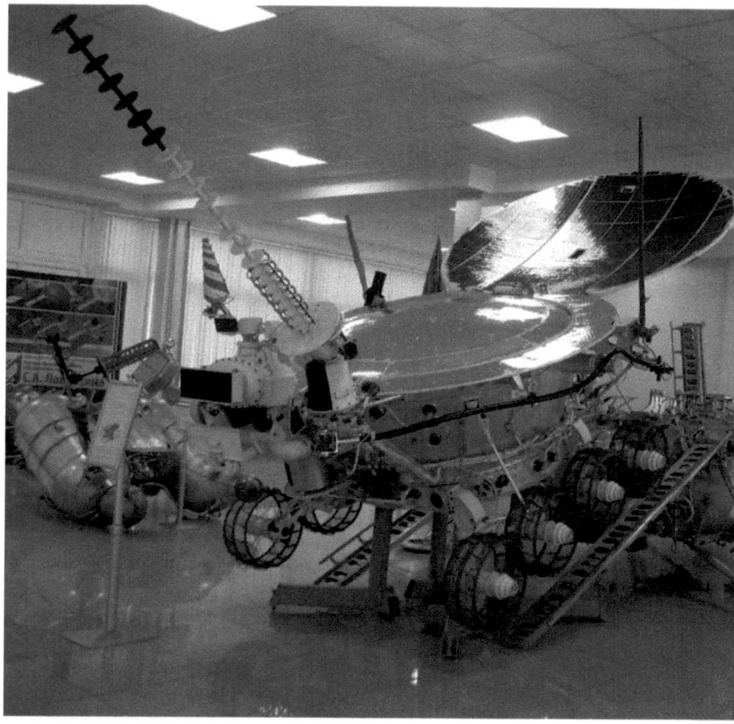

Fig. 3.15 *Lunokhod 3* at the Lavochkin Museum in Moscow (Photo: Zh. F. Rodionova)

Lunokhod 3 was supposed to have flown to the Moon in 1977 on *Luna 25*, but the launch did not take place, and the rover is now in the Lavochkin Museum in Moscow (Fig. 3.15).

References

Abdrakhimov, A. M., Basilevsky, A. T., Head, J.W., & Robinson, M. S. (2011, March 7–11). Luna 17/Lunokhod 1 and Luna 21/Lunokhod 2 landing sites as seen by the Lunokhod and LRO cameras. *Proceedings of the 42nd Lunar and Planetary Science Conference* (p. 2220). The Woodlands, Texas, LPI Contribution no. 1608.

Agamalyan, I. S., Rodionova, Zh. F., Surdin, V. G., & Shevchenko, V.V. (2014). *Fotokarta vidimogo polushariya Luny*. GAISH MGU.

Barabashov, N. N. (1963). *O gornuh porodah, kotorue mogut vhodit' v sostav lunnoiy poverhnosti* (s. 279–283). V knige Novoe o Lune.

Blagg, M., & Müller, K. (1935). *Named lunar formations*. London: Percy Lund, Humphries & Co. Ltd. 196 p.

Bronshten, V. A. (1967). *Karta Luny. Vidimaya storona v teleskopicheskom izobrazhenii. 1:5 000 000*. Sostavitel' I.I. Katyayev, Otvetstvennyy redaktor.

Bugayevskiy, L. M. (1998). *Matematicheskaya kartografiya*. Moskva: Izdatelstvo "Zlatoust".

Chertok, B. E. (1996). *Rakety I lyidi* (448 s). Moskva: Kniga 2.

Gusakova, E., Haase, I., Peters, O., Plescia, J., & Robinson, M. (2013, September 1). Cartography of the Lunokhod-1 landing site and traverse from LRO image and stereo-topographic data. *Planetary and Space Science, 85*, 175–187. ISSN 0032–0633, 10.1016/j.pss.2013.06.002.

Gusakova, E., Karachevtseva, I., Shingareva, K., Oberst, J., Peters, O., Wählisch, M., et al. (2012). *Mapping and GIS-analyses of the Lunokhod-1 landing site*. 43rd Lunar and Planetary Science Conference.

Karachevtseva, I., Oberst, J., Scholten, F., Konopikhin, A., Shingareva, K., Cherepanova, E., et al. (2010). *Space Science Reviews, 150*, 81–124.

Karachevtseva, I., Oberst, J., Scholten, F., Konopikhin, A., Shingareva, K., Cherepanova, E., et al. (2013). Cartography of the Lunokhod-1 landing site and traverse from LRO image and stereo topographic data. *Planetary and Space Science, 85*, 175–187.

Karachevtseva, I. P., Peters, O., Oberst, J., Konopichin, A. A., Shingareva, K. B., Scholten, F., et al. (2011). *The book of abstracts of the 2-nd Moscow solar system symposium (2M-S3)* (pp. 113–114). Space Research Institute (IKI).

Kuiper, G. P. (Ed.). (1960). *Photographic lunar atlas*. Chicago: Chicago University Press.

Lipskiy, Yu. N. (1967). *Fotokarta vidimogo polushariya Luny. 1:5 000 000*. GAISH MGU, TGS SSSR, Nauchnyy rukovoditel'.

Lipskiy, Yu. N. (1968). *Karta Luny. Ekvatorial'naya zona vidimogo polushariya. 1:1000 000*. GAISH MGU, TSNIIGAiK. Nauchnyy rukovoditel'.

Peredvizhnaya laboratoriya na Lune. Lunokhod 1. Barsukov ed. (1971) Tom 1. Izdatelstvo Nauka. Moskva.

Peredvizhnaya laboratoriya na Lune. Lunokhod 1. Barsukov ed. (1978) Tom 2. Izdatelstvo Nauka. Moskva.

Pervuye panoramu lunnoy poverhnosti. (1966) Tom 1. Izdatelstvo Nauka. Moskva.

Pervuye panoramu lunnoy poverhnosti. (1969) Tom 2. Izdatelstvo Nauka. Moskva.

Rodionov, B. N., Nepoklonov, B. V., Kiselev, V. V., Selivanov, A. S., Govorov, V. M., Zasetskiy, V. V., et al. (1973). Topograficheskie S'emki na Lunnoy Poverhnosti s Sovetskih Avtomaticheskih Kosmicheskih Apparatov. *Geodeziya I Kartografiya, 10*, s. 29.

Rodionova, Zh. F. (1969). Luna umen'shennaya v million raz. *Zemlya i vselennaya, 4*, s. 93–96.

Sharonov, V. V. (1963). *Mikrorel'ef Lunnoiy poverhnosti I veroyatnue puti ego obrazovaniya* (s. 283–288). V knige Novoe o Lune.

Shingareva, K. B., & Burba, G. A. (1978). Sistema uslovnykh znakov dlya lunnykh topograficheskikh planov, razrabotannaya po materialam s"yemki «Lunokhoda 1». V kn. Peredvizhnaya laboratoriya na Lune Lunokhod 1. t.2. Nauka, s. 80–91.

Sutinskaya, N. N. (1963). *Meteorno-shlakovaya teoriya lunnoy poverhnosti* (s. 288–292). V knige Novoe o Lune.

Troitskiy, V. S. (1963). *Radioizluchenie Lunu, fizicheskoe sostoyanie i priroda ee poverhnosti* (s. 354–356). V knige Novoe o Lune.

Volkov, N. M. (1964). Vneshniye perspektivnyye proyektsii s pozitivnym izobrazheniyem poverkhnosti zemnogo shara. *Geodeziya i kartografiya, 4*, 70–73.

Chapter 4
Thematic Mapping of the Moon

The history of science is useful if for no other reason than because it saves us the trouble of designing and considering things already discovered and invented by others
F.F. Petrushevskii (1828–1904)

Abstract The development of thematic mapping of the Moon remains a key task in modern lunar studies The first thematic lunar map was constructed in the USSR in 1960. It was The *Map of lunar relief features distinguished by relative degree of preservation and by chronological sequence of formation.* A *Tectonic map of the Moon*, *The Geological map of the Moon*, *Photometric map of the Moon*, a *Map of Albedo* and a *Map of Colour* for the visible hemisphere, a *Polarimetric Atlas of the Moon* were prepared in the USSR. The *Atlas of the Terrestrial Planets and their Satellites* published by MIIGAiK in 1992 includes many thematic maps of the Moon, Earth, Venus, Mercury, Mars and its satellites. GIS software products revolutionised the work of cartographers, providing them with new mapping tools and technologies. Automatic mapping allowed large data volumes to be processed. Recent data from spacecraft surveying the lunar surface have permitted the compilation of a new generation of hypsometric maps and globes of the Moon. Data from the NASA spacecraft *Lunar Reconnaissance Orbiter* were used to produce a *Hypsometric Map of the Moon* at 1:13,000,000 scale.

The idea that the physical parameters of lunar soil can in principle be determined by analyzing the properties of the light scattered by the lunar surface was first voiced at a meeting of the Russian Physical Society presided over by D. I. Mendeleev. On 8th March 1873, the Russian physicist Fyodor Fomich Petrushevskii proposed a plan for the physical study of the lunar surface (Petrushevskii 1873). He developed a unique spectroscope that could be used to compare the spectra of areas of the lunar surface with the spectrum of the entire Moon. Today, we have techniques allowing the chemical and mineralogical composition of any part of the lunar surface to be reconstructed from optical measurements of the Moon and lunar soil samples (Shkuratov 2013).

V. Shevchenko et al., *Lunar and Planetary Cartography in Russia*, Astrophysics and Space Science Library 425, DOI 10.1007/978-3-319-21039-1_4

The development of thematic mapping of the Moon remains a key task in modern lunar studies (Shevchenko 1983). Selenological, gravimetric, hypsometric, structural and morphological, spectral, polarimetric and other techniques of thematic mapping reveal the particularities of the structure and nature of the lunar surface. At the same time, when applied to these maps, cartographic methods are an efficient tool for studying the development of various features and revealing their spatial interrelations and associated phenomena (Rodionova 2010).

In terrestrial cartography, thematic and complex atlases and maps account for about 80 % of the entire cartographic production. The compilation of maps clearly shows that cartographic modelling of the diverse features of the natural environment is an essential component of their investigation and exploration. This experience is, in essence, fully applicable to studies of the nature of the natural satellite of our planet.

The natural environment of the Moon—its *landscape sphere*–has a relatively simple structure. *From above*, the lunar environment is bounded by open space with all the processes that occur in it. Modern data from selenology, selenochemistry and lunar physics provide evidence of active processes that occurred in the lunar lithosphere in the past. Certain activity in the depths of the Moon may also go on in the present epoch as shown, in particular, by seismic studies. The lithosphere can therefore be viewed as the component that shapes the lunar landscape sphere *from below*.

The structure of the cartographic method when applied to lunar studies and its relation to other directions of lunar research can be represented in the form of the following diagram (Fig. 4.1).

The data derived from the results of various studies of the lunar surface and lithosphere are superimposed onto a basemap: either a photomap or a general-purpose schematic selenographic map. The analytical maps based on these data allow the typological classification of individual regions or even of the entire lunar surface. The classification of surface areas with typical properties can be compiled into synthetic maps based on the results of direct studies or the investigation of analogues. These cartographic materials can later be used to perform individual classifications and then to construct a detailed model of the nature of the lunar surface, which, in turn, can be of use when addressing fundamental problems of lunar research, selecting spacecraft landing sites, etc.

The diagram in Fig. 4.2 demonstrates the potential of the Moon's natural and reflected radiation in different spectral regions for imaging with the aim of composing analytical maps containing various properties of the lunar landscape environment.

Remote sensing studies, which provide the basis for the content of various thematic maps, have certain particularities. Observations performed onboard spacecraft allow the entire range of the natural and reflected lunar radiation shown in this figure to be used. At the same time, one must take into account the differences in intensity of these radiations. For example, because of the low intensity of gamma- and X-ray radiation from the lunar surface, it can be usefully resolved only from a sufficiently small distance from the surface: from low lunar orbits with altitudes of about 100 km. As is evident from Fig. 4.2, an analysis of the gamma-ray radiation

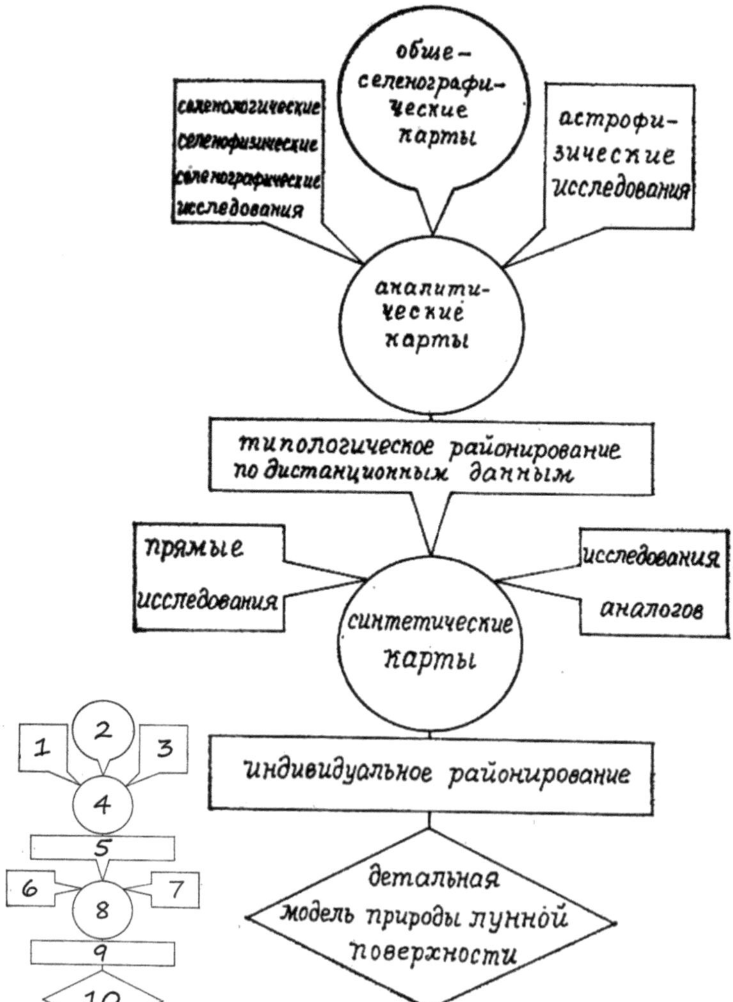

Fig. 4.1 Scheme of the cartographic method of the study of the nature of the lunar surface (1–selenologic, selenophysical, selenographic research; 2–general selenographic maps; 3–astrophysical research; 4–analytical maps; 5–typological zoning from remote-sensing data; 6–direct research; 7–research of analogues; 8–systematic maps; 9–individual zoning; 10–detailed model of the nature of the lunar surface)

from surface rocks provides information about the presence of natural radioactive elements: potassium, thorium and uranium. The X-ray fluorescent radiation from the lunar soil can be used to reliably determine the content of magnesium, aluminium and silicon. Optical imaging of reflected solar radiation of the surface material of the Moon is a universal technique. These data are used to construct photometric,

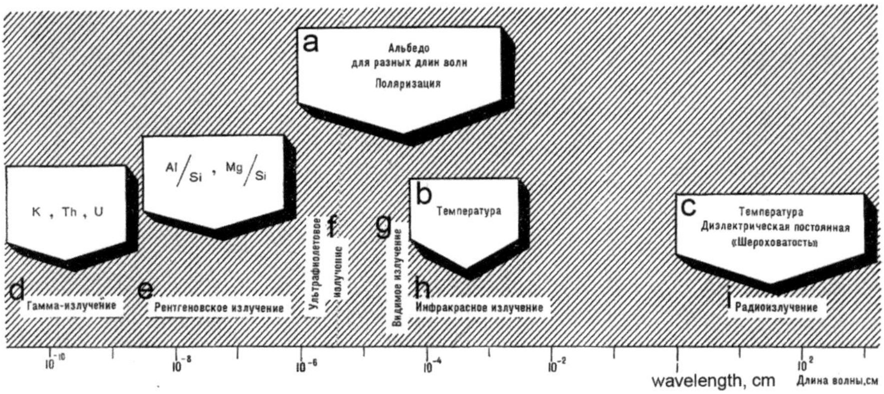

Fig. 4.2 Schematic representation of the use of the Moon's natural and reflected radiation when composing analytical maps about various properties of the lunar surface and lithosphere (a–Albedo for various wavelengths, Polarisation; b–Temperature; c–Temperature, Dielectric constant, Roughness; d–Gamma radiation; e–X-ray radiation; f–UV radiation; g–Visible radiation h–IR radiation; i–Radio emission)

brightness, polarimetric and spectrozonal maps of the lunar surface, which allow deductions concerning the chemical composition and mechanical properties of the surface layer of lunar regolith (Shevchenko 1983).

Infrared imaging is used to determine the brightness temperature of the surface. The patterns of variation of the temperature of the surface layer allow the identification of regions with different thicknesses of the dust cover, exposure of hard rock, etc. Far-infrared and radio observations allow more extensive and elaborate studies of these properties and detailed data to be obtained for a certain depth below the lunar surface.

Figure 4.3 shows a detailed schematic diagram of the classification of thematic lunar maps at different stages of cartographic analysis and synthesis.

The Moon was the first step on the road of exploration of the Solar system by humanity. At the end of the fifties, researchers' interest in the Moon increased in connection with the discovery of some active processes there. It has since become the object of intensive investigation by the latest techniques and capabilities of astronomy, cosmonautics, radiophysics, geochemistry, geophysics, geology and other fields of science and engineering. Even the polarimetry of the light reflected by the lunar surface has acquired a new quality.

The first thematic lunar map constructed in the USSR was the *Map of lunar relief features distinguished by relative degree of preservation and by chronological sequence of formation* (Fig. 4.4). This map represents the main periods of the history of development of the lunar surface (Khabakov 1960). Khabakov used different colours to distinguish ring mountains and linear features of four age categories: very ancient, ancient, young and recent. He proposed dividing lunar surface history into seven main time periods: Recent, Copernican, Oceanic, Ptolemaic, Altaian, Hipparchan (pre-Altai), and Ancient. The map is drawn in telescopic orientation.

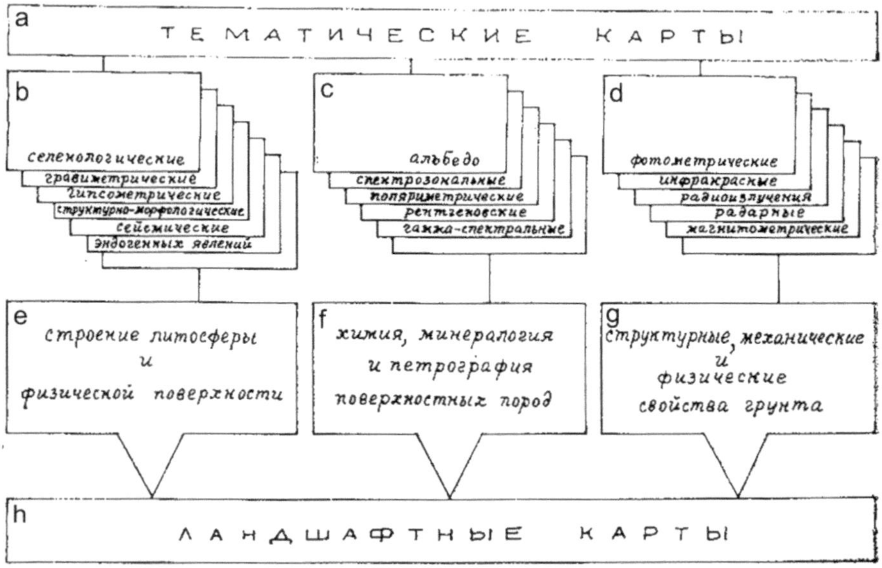

Fig. 4.3 Classification scheme of thematic maps of the Moon (a–Thematic maps; b–selenologic, gravimetric, hypsometric, structural and morphologic, seismic, endogenic phenomena; c–albedo, spectral, polarimetric, X-ray, gamma-spectral; d–photometric, infra-red, radio, radar, magnetometric; e–lithospheric structure and physical surface; f–chemistry, mineralogy and petrography of surface rocks; g–structural, mechanical and physical properties of lunar soil; h–landscape maps)

The Space Research Institute of the Academy of Sciences of the USSR (IKI) prepared a standard base for thematic mapping in the form of 1:5,000,000, 1:10,000,000 and 1:25,000,000 scale blank maps of the two lunar hemispheres, published in 1973 by the Directorate-General for Geodesy and Cartography of the USSR Council of Ministers (GUGK).

Complex tectonic mapping of the Moon should be performed by a large team of experts within the framework of a coordinated programme, and concluded in the form of a series of various maps and atlases providing comprehensive characterisation of an integral group of phenomena (Florenskii et al. 1978).

A *Tectonic map of the Moon* at 1:7,500,000 scale, edited by Yu. Ya. Kuznetsov was compiled on the base of the *Complete Map of the Moon* (1969) in the Laboratory of VNII Zarubezhgeology. Compilers: V.V. Kozlov, E. D. Sulidi-Kondratiev (Tektonicheskaya karta Lunu 1969). The Geological Institute of USSR Academy of Sciences compiled the *Geomorphological Maps of the Moon* at 1:1,400,000 scale and maps of volcanic forms and structural-geological maps at 1:1,000,000 scale (Trifonov and Florenskii 1969; Rodionova 1991). Figure 4.5 shows one such geomorphological map.

The Geological map of the Moon at 1:5,000,000 scale compiled at the Geological Institute of the USSR Academy of Sciences has a detailed legend. The map shows craters—Copernican, Eratosthenian, Archimedean, secondary craters of Mare Orientale, secondary craters of Mare Imbrium, Ptolemean, secondary craters

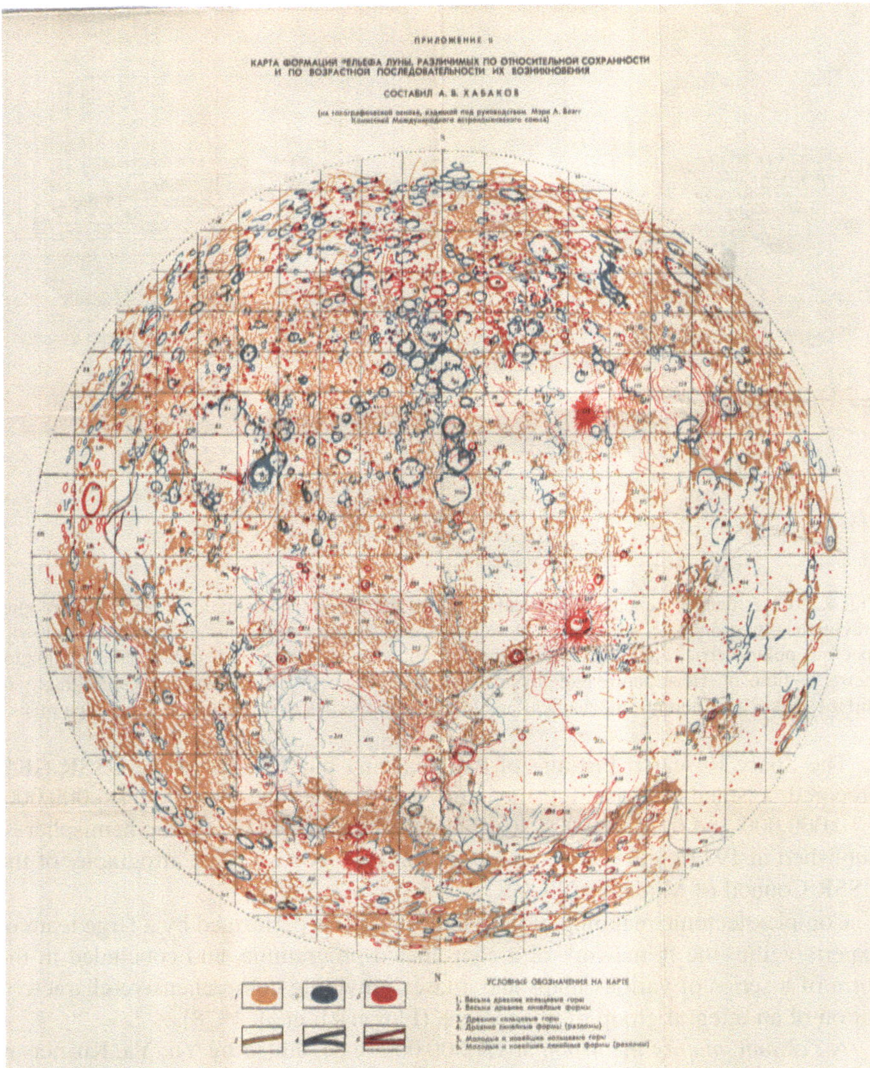

Fig. 4.4 Map of lunar relief features distinguished by relative degree of preservation and by chronological sequence of formation (Khabakov 1960)

of ancient maria and Hipparchan—as well as volcanic features, mare ridges, rilles, ash blankets, mare lavas, pre-mare volcanogenic plains, ejecta of mare depressions and continental structures (Suhanov 1983).

The Astronomical Observatory of Kharkov State University compiled a *Photometric map of the Moon*, a *Map of Albedo* and a *Map of Colour* at 1:5,000,000 scale for the visible hemisphere.

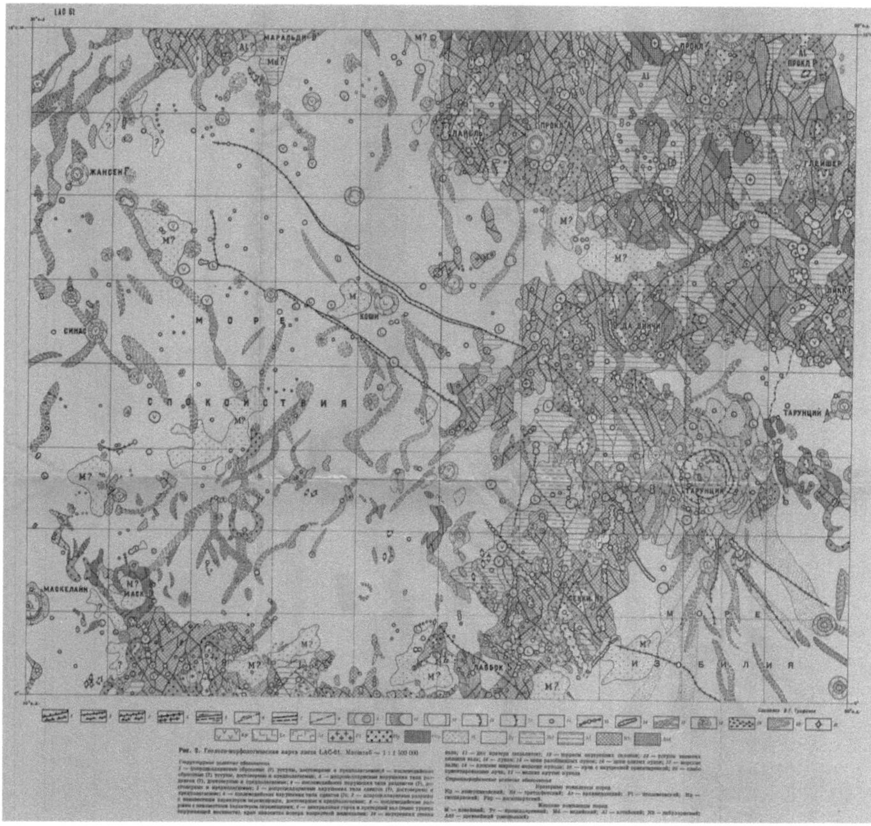

Fig. 4.5 Geomorphological map based on sheet No. 61 LAC

N. N. Evsyukov (Evsyukov 1973) compiled the *Colour Map of the Lunar Nearside* (Fig. 4.6) based on photographs taken at a phase angle of 2° in the 0.62 and 0.38 μm spectral regions. The colour-separating image was obtained by combining the positive red image with the negative blue image with the contrast levels equalised to equal slopes of the characteristic curves. The colour-index contours were obtained using the equidensitometry technique (Shevchenko 1980). All three maps are drawn in orthographic projection.

A *Polarimetric Atlas of the Moon* was prepared by the Abastumani Observatory (Fig. 4.7) (Dzapiashvili and Korol 1982). It was one result of the intensive data collection at Abastumani Astrophysical Observatory in the period of 1971–75 in a process of persistent, pioneering works on studying the lunar surface with the electropolarimetric method (Dzapiashvili and Korol 1978), a traditional technique for this observatory in investigating celestial objects. A series of observations for 46 lunar phases was carried out between 1971 and 1975. Ninety polarimetric records were obtained on magnetic tape, with observations at each phase usually being duplicated. The records of 21 lunar phases were used for the *Polarimetric*

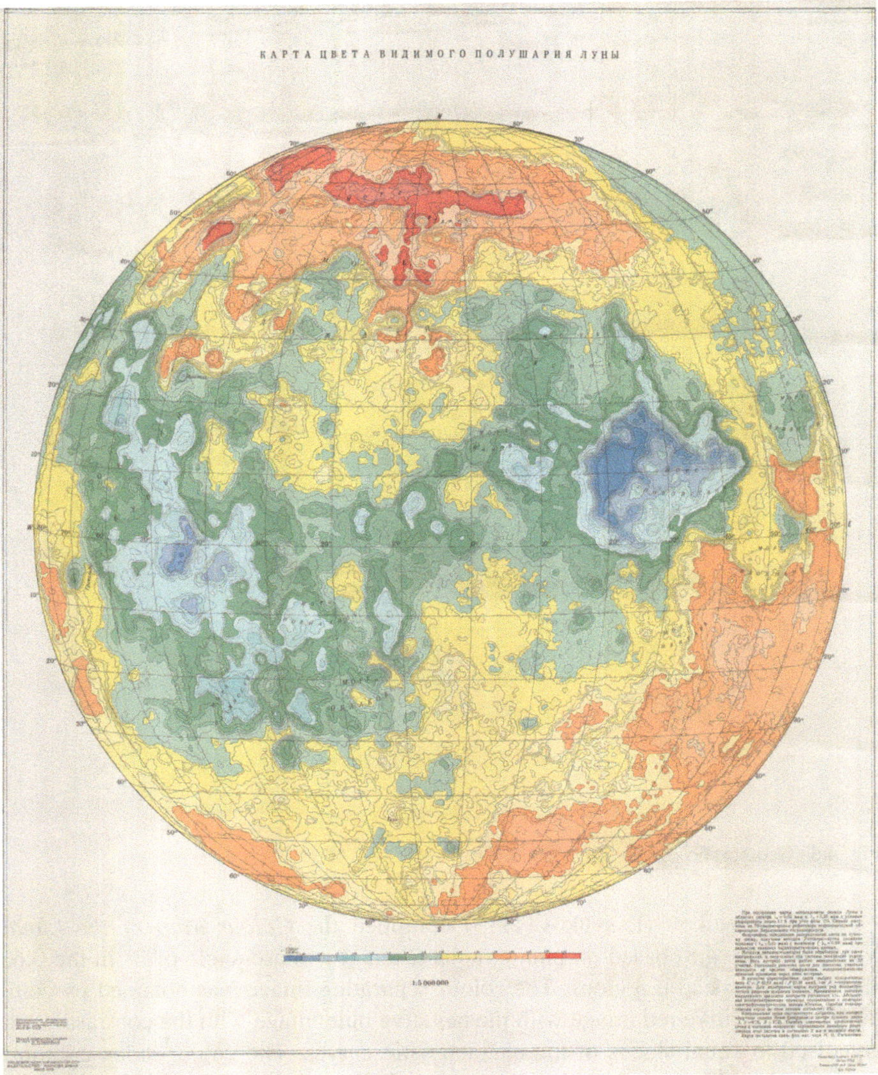

Fig. 4.6 Colour map of the lunar nearside

Atlas of the Moon. The polarimetric maps in the atlas are arranged in the order of increasing values of the lunar phase angle, i.e., the angle subtended at the centre of the lunar disc by the directions to the Sun and the Earth. The phase angle is conventionally regarded to be negative before a full Moon and positive after.

In the upper left corners of the maps, there are frame numbers corresponding to the chronological sequence in which the images were obtained, and in the upper middle there are the original numbers of the maps according to their arrangement in the atlas. In the upper right corner, there is a date and average time of imaging in UT

Fig. 4.7 Polarimetric atlas of the Moon with sample page

as well as the lunar phase angles corresponding at that time (Fig. 4.8). The maps were compiled in perspective orthographic projections at 1:15,000,000 scale. The numbers beside the legend denote the colours in the maps corresponding to the degrees of polarisation of the reflected light.

The *Atlas of the Terrestrial Planets and their Satellites* (Atlas Planet Zemnoy Gruppu i ih Sputnikov 1992) published by Moscow State Institute of Geodesy and Cartography (MIIGAiK) in 1992 includes many thematic maps of the Moon, Earth, Venus, Mercury, Mars and its satellites. The atlas was a collaborative project with the participation of many institutes and organisations. The 20 sections of the atlas cover the following subjects: history of mapping of terrestrial planets and their satellites, maps of planetary surfaces, blank maps, hypsometric maps and cartometry studies, maps of physical surface properties, geophysical and geomorphological maps, unmanned space probes and manned spacecraft flights, the state of photographic, geodetic and cartographic exploration of celestial bodies. The atlas consists of 208 pages, most of them with maps. Here, we mention some of the thematic maps compiled for the Moon.

Figure 4.9 shows the *Geomorphological Map of the Moon* at 1:25,000,000 scale. Different colours are used to distinguish highland cratered surfaces, intensely cratered highland surfaces, cordilleras and mountainous areas bounding large basins and mare surfaces. Figure 4.10 shows a *Map of the Lunar Nearside Distribution of Non-stationary Phenomena* observed over the past 400 years.

Different symbols in this map show different types of non-stationary phenomena such as glows and bright spots; darkenings or haze; brightness changes; red spots; color

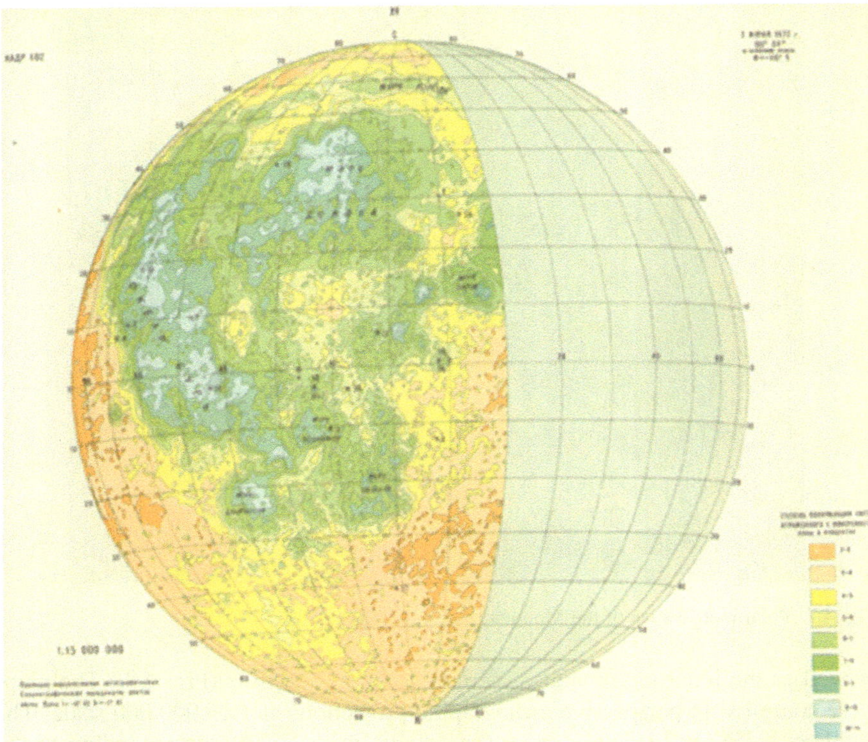

Fig. 4.8 Polarimetric map of the atlas (frame no. 602)

changes and dark spots. Note that 1 mm of the bar height corresponds to one event. Because of the large number of events in Aristarchus and Alphonsus craters, which are indicated by an asterisk in the map, their corresponding symbols are shown separately.

The *Map of the Distribution of Thermal Anomalies on the Nearside of the Moon* (Fig. 4.11) shows the thermal emission features of the lunar surface at wavelengths 10–12 μm.

The map shows the spatial arrangement of thermal anomalies identified from lunar surface temperature measurements during eclipses. Areas where the surface temperature exceeds the temperature of the surrounding region are indicated. Special symbols are used to distinguish four groups of thermal anomalies. Extended regions with anomalous surface temperature values in lunar maria are also shown. The map gives an insight into the emission of the lunar surface.

GIS software products revolutionised the work of cartographers, providing them with new mapping tools and technologies. Automatic mapping allowed large data volumes to be processed and the results presented within a short time (Lazarev and Rodionova 2007).

The publication of the *Relief Map of Lunar Polar Regions* (Fig. 4.12) was timed to commemorate the 100th anniversary of the birth of Yury Naumovich Lipsky,

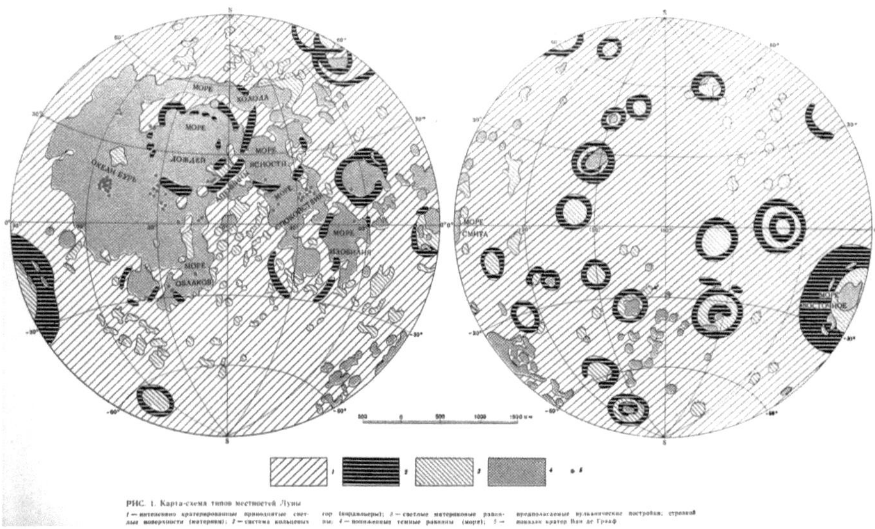

Fig. 4.9 Geomorphological map of the Moon from the atlas of the terrestrial planets and their satellites

who headed the Department of Lunar and Planetary Research of GAISh until 1978. The *Complete Map of the Moon* on nine sheets (third edition) and a reduced scale version of it—the brainchild of Yury Naumovich—were published posthumously. The map had a “white spot” in the south polar region, south of Cabeus crater, which long remained unphotographed. Only in the twenty-first century, ground-based radar imaging from Arecibo radio observatory combined with mapping from ESA’s *SMART 1*, the Japanese *Kaguya*, the Chinese *Chang’e 1*, and the Indian *Chandrayaan 1* space probes made it possible not only to image this region (Fig. 4.13) but also determine the altitude distribution across the entire lunar surface.

The *Relief Map of Lunar Polar Regions* (Fig. 4.12), compiled at GAISh based on the analysis of more than 2 million altitude measurements obtained by the Japanese *Kaguya* spacecraft, shows 500 m height intervals in different colours. The heights are measured from the mean level of the lunar surface—1737.4 km, adopted by the IAU. Shades of green indicate areas below this level, and shades of red areas above it. Places of particular interest are marked with numerical height values, e.g., the floor of Shackleton Crater is −2.8 km at its deepest, and a point on the rim is marked with +1.2 km. The polar elevation data were processed in particular to study the permanently shadowed areas of craters, the sites of possible *cold traps* (Fig. 4.14).

Recent data from spacecraft surveying the lunar surface have permitted the compilation of a new generation of hypsometric maps and globes of the Moon. Data from the NASA spacecraft *Lunar Reconnaissance Orbiter* were used to produce a *Hypsometric Map of the Moon* at 1:13,000,000 scale. The map was constructed in ESRI’s ArcGIS 10.1 using the DEM at 64 pix/degree (~0.5 km/pix). The map (Fig. 4.15) was made in equal area azimuthal Lambert projection for both the near- and farsides. For polar regions out to 60° latitude, we used a polar stereographic Lambert projection.

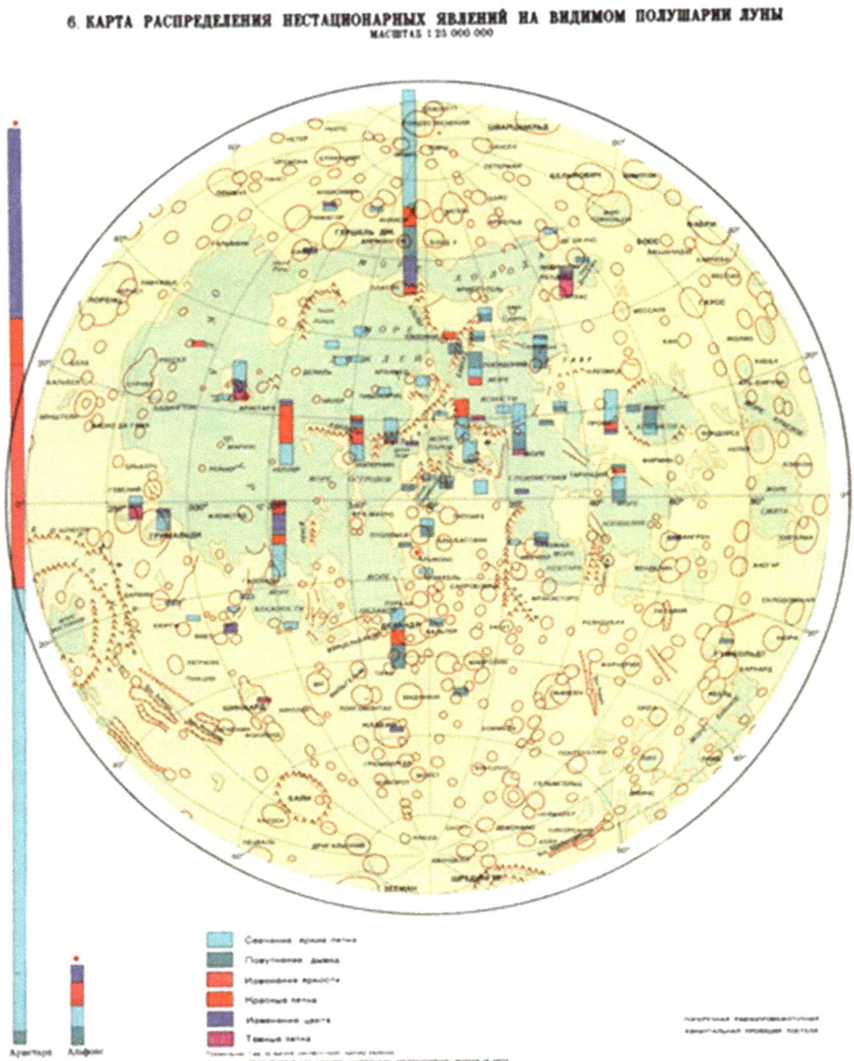

Fig. 4.10 Map of the lunar nearside distribution of non-stationary phenomena from the atlas

The height scale, intervals and colour representation were developed by E. N. Grishakinaya. The Spatial Analyst module was used to calculate and generalise contours in a separate vector layer and to generate the hillshade using solar incidence angles of 315° and 45° and a z-factor of 5. Surface features were labelled according to the IAU nomenclature database. Contours were labelled to make the map as informative as possible while remaining uncluttered. Choices in the number of named features and of lettering styles and sizes were made with the same aim (Grisgakina et al. 2015)

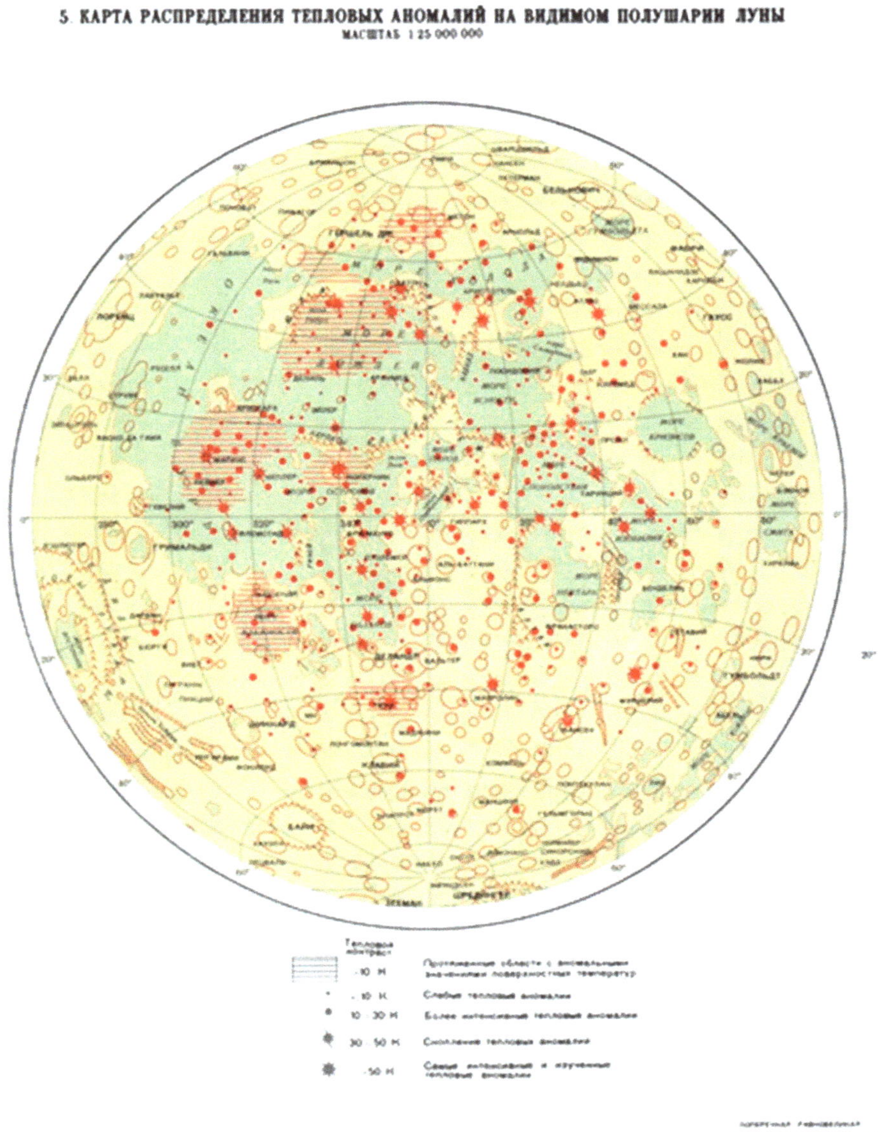

Fig. 4.11 Map of the distribution of thermal anomalies on the nearside of the Moon from the atlas

A *Hypsometric Globe of the Moon* (Fig. 4.16) was produced at IKI by J.A. Brekhovskikh on the basis of the *Hypsometric Map of the Moon*, reprojecting the map into north and south hemispheres appropriate for the thermoplastic manufacturing process (Brekhovskikh et al. 2014).

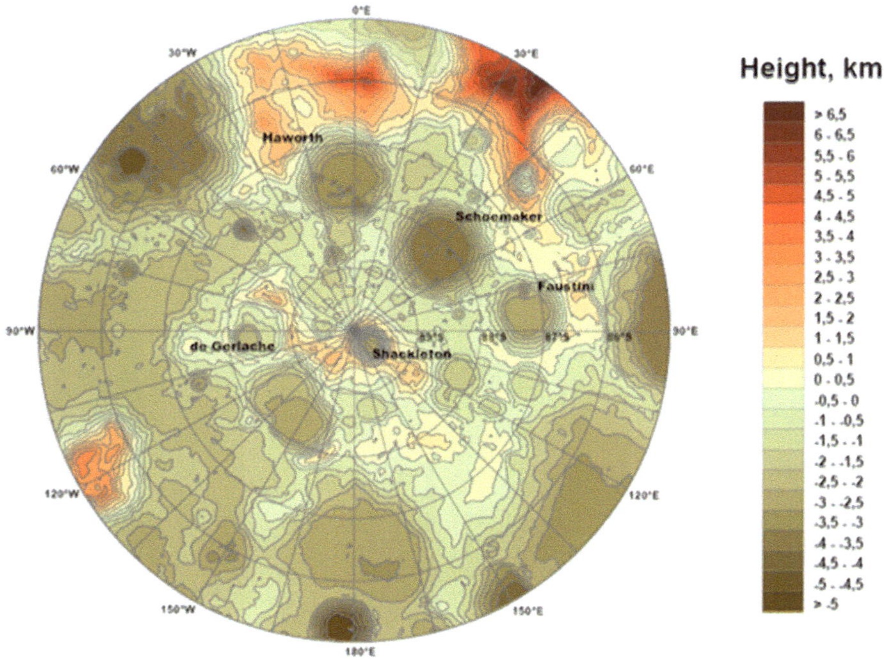

Fig. 4.12 *Relief Map of lunar polar regions* compiled at GAISh based on the data obtained by the Japanese *Kaguya* space probe

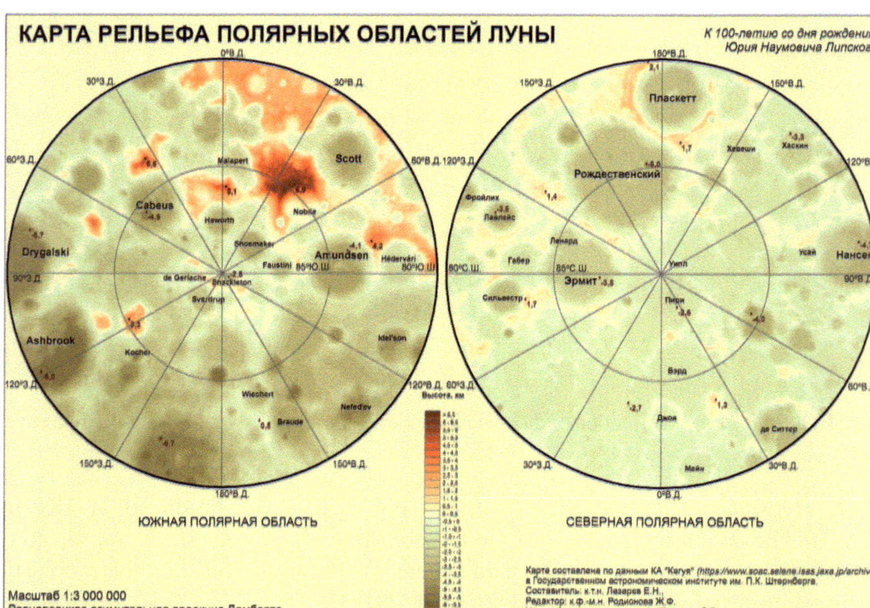

Fig. 4.13 Fragments of the South and North polar areas shown on the *Complete Map of the Moon* (1979)

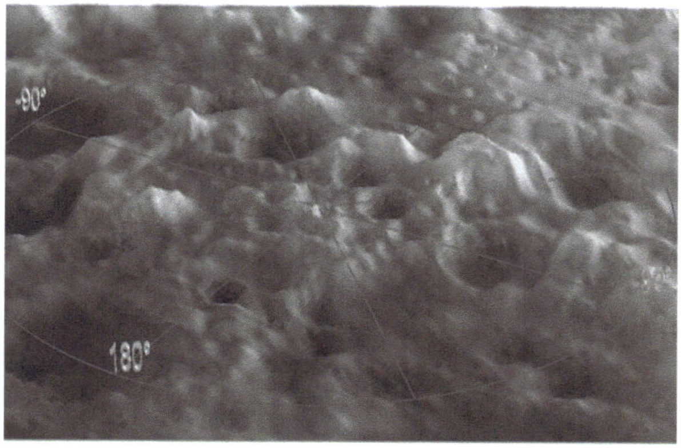

Fig. 4.14 Digital elevation model of the south polar region based on *Kaguya* data

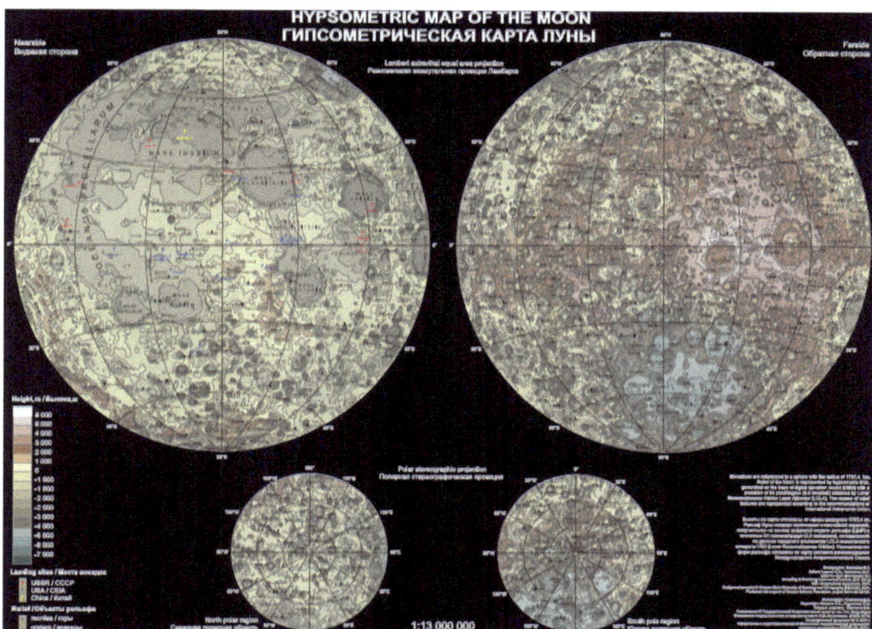

Fig. 4.15 Hypsometric map of the Moon

Fig. 4.16 Hypsometric globe of the Moon (Brekhovskikh et al. 2014)

References

Atlas Planet Zemnoy Gruppu i ih Sputnikov (1992). Izdatelstvo MIIGAiK (208 s).

Brekhovskikh, J., Grisgakina, E., Lazarev, E., Rodionova, Z., & Shevchenko, V. (2014). *Gipsometricheskiy globus Luny*. Moskva: IKI, GAISh.

Dzapiashvili, V. P., & Korol, A. N. (1978). Nauchnue Informatsii Astronomicheskogo Soveta AN SSSR (seria Astrofizika). *Izdatelstvo AN SSSR, 45*, 205.

Dzapiashvili, V. P., & Korol, A. N. (1982). *Polarimetricheskiy Atlas Lunu* (45p.). Tbilisi.

Evsyukov, N. N. (1973). Statisticheskoe Raspredelenie Albedo po Lunnomy Disky. *Astronomicheskiy Zhurnal, 50*(1).

Florenskii K. P., Bazilevskii A. T., & Bobina, N. N. (1978). *Tematicheskoe kartografirovanie Lunu*. IV kn. Kartografirovanie Lunu I Marsa (pp. 66–72).

Grisgakina, E., Lazarev, E., Rodionova, Zh., & Shevchenko, V. (2015). *Gipsometricheskay karta Luny*. Moskva: GAISh MGU.

Khabakov, A. V. (1960). *Harakternue Osobennosti Reliefa Lunu. Osnovnue problemu Genezisa i Posledovatel'nosti Razvitiya Lunnuh Formatsiy*. V kn. "Luna" pod redaktsiey Markova A. V (s. 241–295).

Lazarev, E., & Rodionova, Zh. (2007). The lunar subpolar relief map: The ways and techniques of compiling and using. *Proceedings of XXIII International Cartographic Conference* (s. 371).

Petrushevskii, F. F. (1873). Plan Fizicheskogo issledovaniya Lunnoy Poverhnosti. *Zhurnal Russkogo Fiziko-Himicheskogo Obschestva,-T. 5*, vupusk 5, s. 219–238.

Rodionova, Zh. F. (1991). Space achievements and maps and globes of planets. *Geodesy and Cartography, 7*, 15–22.

Rodionova, Zh. F. (2010). *Luna (The Moon)*. Department of Lunar and Planetary Research of Sternberg Astronomical Institute: Lecture about the Moon. http://selena.sai.msu.ru/Home/SolarSystem/moon/moon.htm
Shevchenko, V. V. (1980). *Sovremennaya selenografiya* (287 pp.).
Shevchenko, V. V. (1983). Problemu Tematicheskogo Kartografirovaniya Lunu. Izv. vuzov. *Geodeziya i Aerofotos'emka, 1*, 62–70.
Shkuratov, Yu. G. (2013). *Khozhdenie v nauku*. Khar'kov (275 pp.).
Suhanov, A. L. (1983). Geologicheskoe Kartirovanie Lunu. Izv. vuzov. *Geodeziya i Aerofotos'emka, 1*, 73–86.
Tektonicheskaya karta Lunu. (1969). *Redaktor Kuznetsov, VNII Zarubezhgeology*. Moskva.
Trifonov, V. G., & Florenskii, P. V. (1969). *Problemu geologii Luny* (281 pp.).

Chapter 5
Essays on Lunar Toponymy. Events and People Reflected in the Names on Lunar Maps

Abstract The individual names of details of relief are important part of the cartographic information. The history of the names shows us cultural background of the period when they were given. The appearance of names of different forms of relief is intimately connected with the process of compilation of maps of planetary bodies, and as such it is logical here to examine the history of lunar cartography. The lunar craters are given the names of astronomers, primarily researchers of the Moon, and the names of prominent scientists and explorers. The names of classes of morphologic structures on the Moon, such as seas and lakes, are constructed either from names of phenomena connected with the weather or with emotional states. Some formations have names which are analogous with number of features on the Earth. A new epoch of development of lunar toponymy began almost simultaneously with the start of the space age. In hundreds of years, our descendants will look back on our time, understanding it, in part, from this source of information.

A significant part of the information on a map is the set of individual names of details of relief which allow the reader to identify a feature of interest. Each such name is a short code for the location of an object, characterising its structure (nomenclature), and also having a cultural and historical meaning. In most cases, the name is sufficient for a specialist to describe exactly the location of some or other object that he has in mind. Examining the history of the names makes it possible to reconstruct the circumstances and cultural background of the period when they were given and its level of scientific and technical development. At the present time, there are around 7300 named features across 37 bodies of the Solar System.

According to the classical definition, toponymy is a complex scientific discipline at the juncture of three fields of knowledge, using the methods and data of geography, history and linguistics. In this case, we understand geography in the wide sense of the science to study the surfaces of bodies of the Solar System. Its most frequent analogues are selenography (the science of the study of the lunar surface), areography (the science of the study of the surface of Mars), etc. Toponymy itself is a science, the goal of which is to study the names of surface features and clarify their origins and significance. It has become generally accepted to use the term *toponymy* (from the Greek *topos*—"place" and *onoma*—"name") to describe the naming of surface features on planetary bodies. Consequently, the

V. Shevchenko et al., *Lunar and Planetary Cartography in Russia*, Astrophysics and Space Science Library 425, DOI 10.1007/978-3-319-21039-1_5

collection of named objects on one or other body is called the toponymy of Mercury, Venus, Mars, the Moon, etc.

The toponymy with the longest history is that of the Moon, and from this example we can follow the evolution of planetary toponymy and typical problems that have been encountered by researchers in creating naming systems (or nomenclature) for planetary surfaces.

A tradition gradually emerged by which the toponymy of each of the bodies of the Solar System is constructed according to specific rules. The basis of the current planetary toponymy was laid down in the seventeenth century in the process of creating a naming system for features of the lunar surface. In the present day, surface features of Mercury are given the names of eminent cultural individuals (writers, artists or musicians), famous research ships, the word for Mercury in different languages and the names of Earth-based radio observatories. On Venus, features are named in honor of prominent women, with girls' first names, with the names of goddesses and heroines of Mythology and also with words for Venus in different languages. Features on Mars are named in memory of distinguished researchers of the planet, authors of science fiction about travel to Mars, with names from the ancient geography of the Mediterranean and with the words for Mars in different languages. The multitude of mythological names from the folklore of different peoples of the globe are used to name features on the surfaces of satellites of the giant planets and those asteroids and cometary nuclei which have been imaged at high resolution by spacecraft.

History of the Compilation of Lunar Toponymy

The appearance of names of different forms of relief is intimately connected with the process of compilation of maps of planetary bodies, and as such it is logical here to examine the history of lunar cartography. The first map of the visible hemisphere of the Moon was created even before the first telescopes by the famous British physicist, the discoverer of the Earth's magnetism, William Gilbert. This document first appeared in 1651 although, by the estimate of historians, it was compiled not long before his death in 1603 (Kopal and Carder 1974). The drawing was a schematic outline of the dark and light regions of the Moon observed with the unaided eye. Gilbert's map included just over 10 names which were not retained in subsequent works. However, this first attempt of creating a system of lunar names formed the basis of some traditions of toponymy. First of all, sketching the large features of the lunar surface, Gilbert divided it into seas and continents by analogy with the surface of the Earth. Interestingly, he classified the dark regions as continents, which in later lunar cartography came to be called seas, but as a whole this tradition was kept and continued. On Gilbert's map, first appeared terms such as *cape* and *bay*. Finally, in the spirit of those times, all the names were given in Latin, and this tradition has been maintained ever since.

Fig. 5.1 Grimaldi's map, published in the work of Riccioli *Almagestum Novum* (1651)

The first depictions of the visible hemisphere of the Moon made with the aid of a telescope appeared at almost the same time in 1609 in England and in Italy. The more detailed and, from the modern point of view, more realistic representation was made by the English astronomer, Thomas Harriot. The drawings of Galileo Galilei had a more illustrative character. These depictions did not contain any feature names, and neither did several which followed from authors of the first half of the seventeenth century.

Detailed and unrelated systems of lunar feature names appeared on the maps of Michael Florent van Langren (1645), Johannes Hevelius (1647) and Francesco Maria Grimaldi (1651). The map compiled by Grimaldi (Fig. 5.1) was published in a broad work by Giovanni Batistta Riccioli, professor of Philosophy, Theology and Astronomy in Bologna, and his is traditionally considered to be the first elaborated lunar toponymy.

Many names used on Grimaldi's map are still used today, as are the structural principles of toponymy he introduced. In particular, this relates to the tradition to name lunar craters after prominent scientists. Riccioli named craters after eminent contemporaries.

On modern maps, there are several names which were given to lunar features by Hevelius. These are mountain ranges named after terrestrial mountains, such as the Alps, Caucasus, Altai, Carpathians, Apennines and Pyrenees.

Soon after the formation of the International Astronomical Union (IAU), this community of astronomers took on the responsibility of creating and curating planetary toponymy. "Named Lunar formations" by Blagg and Müller (1935) was the first systematic listing of lunar nomenclature. In 1935, the IAU approved this list of about 600 names of objects on the visible side of the Moon: the first to receive an official international status.

The traditions laid down in the works of Riccioli and Hevelius have changed somewhat. According to the rules adopted by the IAU, the names of lunar features are given in honor of significant figures of science and technology after their deaths. Furthermore, the relevant committee of the IAU does not consider proposals earlier than 3 years after the death of the person suggested for commemoration. Thus, a high level of responsibility is ensured for such decisions, for the names which have survived from Riccioli's times have existed already over the course of centuries. Those names which are placed on lunar maps in our times will likewise remain in the history of humanity forever.

According to the established tradition, lunar craters are given the names of astronomers, primarily researchers of the Moon, and the names of prominent scientists and explorers. The names of morphologic classes, such as seas and lakes, are constructed either from names of phenomena connected with the weather or with emotional states. Some formations, such as mountain chains, have names which are analogous with the same features on the Earth.

The development of space research required new efforts. In 1973, the IAU had already formed six working groups of experts who worked on the development of the toponymy of the Moon, Mercury, Venus, Mars and the satellites of the giant planets, asteroids, and comets. From 1978 to 2010, the expert group for the toponymy of the Moon (the Lunar Task Group) was headed by Vladislav Shevchenko. The chairmen of the expert groups, as well as several specialists on names from different countries, make up the International Astronomical Union Working Group for Planetary System Nomenclature (IAU-WGPSN).

The working group takes final decisions on all received proposals and brings the corresponding list for affirmation by the General Assembly of the IAU. The Latin names are considered the official international designations of the objects and are contained in the Gazetteer of Planetary Nomenclature on the website http://planetarynames.wr.usgs.gov. This resource contains the full nomenclature of planets, satellites, asteroids, comets, maps of the feature locations as well as information about the origin of every one of the several thousand names.

A new epoch of development of lunar toponymy began almost simultaneously with the start of the space age in the history of humanity. The first images of the far side of the Moon, which were received as a result of the flight of Luna-3 in 1959, formed the basis for the compilation of the first map of the lunar farside and the first globe of the Moon, and raised the need for an expanded system of lunar names.

The study of the first images of the lunar farside was made by leading specialists of lunar planetary science under the leadership of Yurii Naumovich Lipsky (Fig. 5.2). This group put together the first set of names for the farside of the Moon. After a confirmation by a special committee of the Academy of Sciences chaired by M. B. Keldysh which included the most prominent academics of our country, such as S. P. Korolyov and V. P. Glushko, the list was sent to the International Astronomical Union.

The 11th General Assembly of the IAU took place in 1961 in Berkeley (US). Resolution Number 2 of Committee 16 confirmed the first 18 names of features of lunar farside. Table 5.1 is a photocopy of pages of material from the IAU 11th General Assembly, which shows the list.

Figure 5.3 shows the scheme of imaged regions of the near and far hemispheres of the Moon, published in the same materials.

The discussion and acceptance of these new lunar names laid down a new basis, from which all subsequent proposals followed. Among these new rules was the principle that, whichever the nation behind of a particular achievement in space, the list of names for commemoration should be compiled on an international basis. Thus, among the names of prominent figures of soviet science and engineering, such as Lomonosov, Lobachevskii. Tsiolkovskii and Kurchatov, on the farside of the Moon the names of Giordano Bruno, Joliot Curie, Pasteur, Tsu Chung-Chi, Maxwell, Edison and others are also remembered.

In the process of further research new circumstances arose, relating to phenomena and features which previously were not understood. In interpreting the images,

Fig. 5.2 The first researcher of the lunar farside Yurii Naumovich Lipsky (1909–1978). Also shown is the first lunar globe

Table 5.1 Photocopy of pages of material from the IAU 11th General Assembly

		Coordonnées sur la carte	
Désignation	n° du catalogue	Longitude°	Latitude°
Tsu Chung-Chi	1	141	+18
Kurchatov	2	144	+32
Jules Verne	5	151	−37
Mendeleev	10	167	−2
Popov	71	99	+14
Hertz	74	101	+11
Edison	83	100	+24
Lobachevsky	88	112	+9
Pasteur	95	111	−10
Sklodowska Curie	112	102	−23
Tsiolkovsky	151	131	−22
Lomonosov	177	99	+28
Joliot Curie	183	93	+25
Maxwell	185	99	+30
Giordano Bruno	208	103	+36
Montes Sovietici	103	(de 111	+19
		(à 124	−5
Mare Ingenii	150		
Mare Moscoviense	152	149	+27

the bright band at the center of the lunar disk, as seen from the spacecraft, was taken to be an extended mountain ridge. This feature was given the name *Montes Sovietici* (Fig. 5.3). The interpretation raised no objections in the IAU at that time and was confirmed. However, when further images of the farside were made later with higher resolution, it turned out that the feature was actually an extended light ray from a crater and the IAU removed the previous name at the 14th General Assembly, which took place in Brighton in 1970.

In another case, researches showed excessive caution, as a result of which the discovery of the largest ring structure of the solar system was only made many years later. In the southeastern part of the limb of the lunar disk seen from the spacecraft (Chap. 1, Fig. 1.4), a broad region of uneven dark albedo was discovered. The interpreters of these images took the structure to be a lunar mare, giving it the name Mare Ingenii, according to the rules of the IAU (Fig. 5.3).

In recent times, images of the distant hemisphere of the Moon have shown that the giant depression, which is still known contingently as the South Pole—Aitken basin, really has a lower albedo than the surrounding highlands (Fig. 5.4). Inside the ring structure can be seen a relatively small formation of mare to which the name Mare Ingenii was transferred (Shevchenko 2009).

After new spacecraft had been sent to the Moon (*Zond-3* and a series of *Lunar Orbiters*) the IAU, at its 14th General Assembly in 1970, confirmed a list of nearly

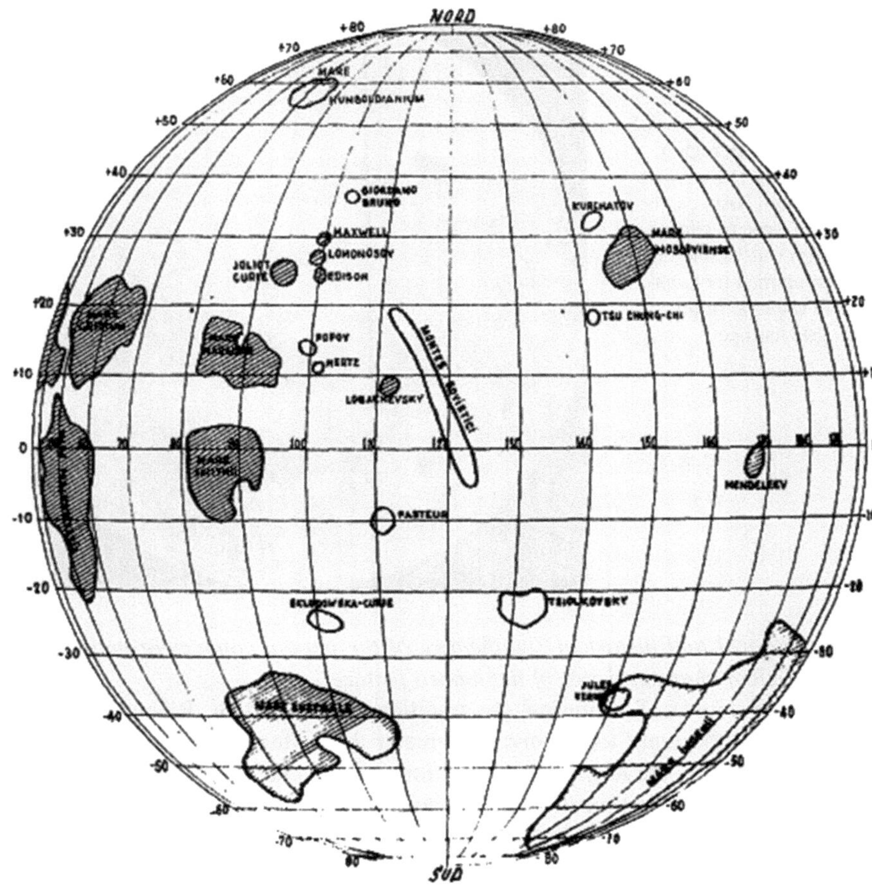

Fig. 5.3 Scheme of locations of features on the farside of the Moon, which were given names by the 11th General Assembly of the IAU, as shown in Table 5.1. On the West (*left*) side of the diagram, structures can be seen from the visible hemisphere which were included in the region of imaging to connect the cartographic coordinates with the territory of the farside

seven hundred names for the farside of the Moon. Thus, the lunar toponymy began to approach 1500 names.

As mentioned above, the creation of the lunar toponymy became the basis for the elaboration of the general methods and principles of planetary toponymy as a scientific field. We can follow this process through several examples.

The contemporary nomenclature of lunar features comprises hundreds of names identifying eighteen morphologic classes of features. The majority of the names refer to craters, including 1521 individual names and 7056 which repeat the main name with an additional identifying letter of the Latin alphabet. Altogether, there is one crater name for every 61 km^2 of the lunar surface (Shevchenko et al. 2009).

Fig. 5.4 Mosaic showing the albedo distribution on the lunar farside based on images from *Clementine* (1994). In the lower part of the disc, the dark region corresponding to the South Pole—Aitken basin can be seen. The western part of the region was visible in images returned by *Luna-3* in 1959 (*Source*: http://nssds.gstc.nasa.gov)

The cultural and historical significance of the person commemorated and the morphological characteristics of the named feature.

Over the course of centuries, the tradition established by Riccioli has been maintained in the lunar toponymy: the greater the cultural and historical significance of the figure, the larger the surface formation named in his honour.

Of course, for a long time the right of choice belonged to the compilers of a map. A good example is Grimaldi's map with the names which are traditionally ascribed to Riccioli. Figure 5.5 shows a fragment of this map. It is worth noting that the largest craters in this region of the visible lunar hemisphere carry the names of Riccioli himself (1), the compiler of the map, Grimaldi (2) and their contemporary, the lunar cartographer Hevelius (3). In this context, we see that the less significant figures are the great astronomers Galileo (4) and Kepler (5), who eventually occupied a more significant place in history of humanity.

If we remember the historic context of the compilation of this map, it becomes clear that the point is not simply Riccioli's mania of self-importance. It is known that in 1616, eleven prominent catholic theologians examined the works of Nicolaus Copernicus and concluded their falsehood. The consequence of this decision was the summoning of Galileo from Florence to Rome with the demand to cease his heretical propaganda about the structure of the World. In 1632, Galileo's book *Dialogue Concerning the Two Chief World Systems* appeared, in which he spoke of the incorrectness of the views of Aristotle and Ptolemy. This time Galileo was summoned to Rome to a court of the Inquisition. The proceedings stretched from April until June 1633. To save his life, he was forced to acknowledge the falsehood of the heliocentric world system and, on the 22nd of June 1633, to read out the proffered text of repudiation on his knees. In his final years until his

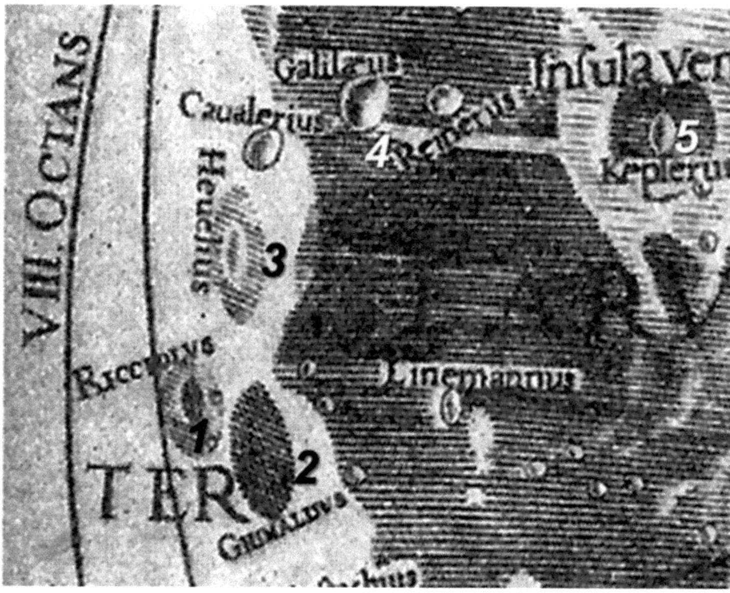

Fig. 5.5 Fragment of Grimaldi's map with the names proposed by Riccioli

death in 1642, Galileo remained under house arrest and under the eyes of the Inquisition.

Returning to Riccioli, it is worth noting that being not only a physicist and an astronomer but also a professor of philosophy and theology, in 1614 he entered the jesuit order. He likewise rejected the heliocentric world system of Copernicus. In his main scientific work, *Almagestum Novum*, Riccioli published the protocol of the proceedings against Galileo, including his text of repudiation. In this respect, Riccioli's placement on the Moon of the names of supporters of the heliocentric world system, Galileo, Kepler and Copernicus himself, can be considered a highly courageous act. All these names have been preserved untouched for centuries in the lunar toponymy.

At the end of the 1960s, when new surface features on the farside of the Moon were being named as a consequence of spacecraft images, it was proposed to transfer Galileo's name to one of the largest craters close to the South Pole. However, according to the rules of the IAU, it is forbidden to move a name from one feature to another, so Galileo's name remained on the small crater where it was placed by Riccioli.

One of the last cases when the size of the structure was related to the historical significance of the figure after whom it was named was the appearance on the farside of the Moon of the large ring structure, Korolev.

This structure was discovered on images received from *Zond-3* (1965), during the first survey of the hemisphere of the Moon not visible from Earth (Fig. 5.6).

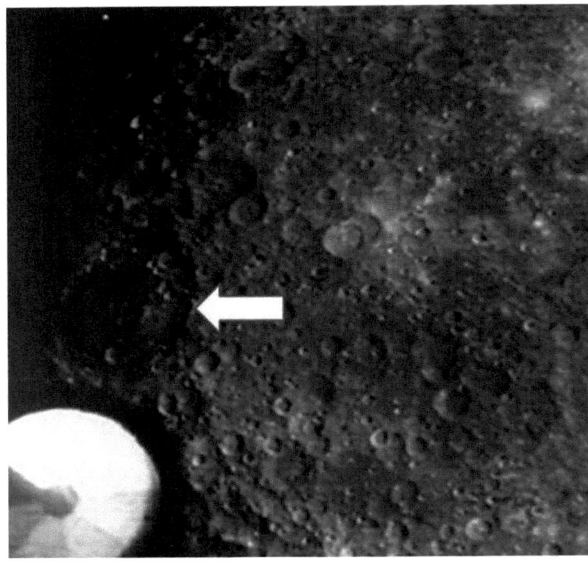

Fig. 5.6 Fragment of an image obtained by *Zond-3*. The *arrow* indicates the ring structure of diameter 423 km which received the name Korolev (Rodionova et al. 1987)

The launch of *Zond-3* to the Moon was the last lunar space experiment carried out under the direction of the prominent scientist and engineer Sergei Pavlovich Korolev (Fig. 5.7). *Zond-3* was an autonomous interplanetary spacecraft, closely analogous to those which were later sent to Mars and to Venus. These flights were also technological tests of the spacecraft systems: in particular, the transmission of the lunar images was made when the spacecraft had reached a distance of 2.2 million kilometres from the Earth (about 6 times the Earth–Moon distance).

Less than a year after the successful flight of *Zond-3*, the chief engineer of the Special Engineering Bureau #1 (OKB-1, today known as the S. P. Korolev Rocket and Space Corporation *Energia*), Academician Korolev died unexpectedly.

At the time of the flight of *Zond-3*, Korolev followed all the transmissions from the spacecraft which were analysed directly and urgently in OKB-1 by a group of scientists led by Lipsky (the author of this book, V. V. Shevchenko was a member of this group). Korolev was one of the first to see the newly discovered ring structure on the farside of the Moon (Fig. 5.6). Consequently, it was this structure that was given his name by decision of the 14th General Assembly of the IAU (1970).

The next problem before the compilers of lunar toponymy in the 70s and 80s was the giving of personal names to large structures on the surface of Earth's natural satellite. As mentioned earlier, the largest circular structure in the solar system remains unnamed to this day, referred to by the contingent name South Pole—Aitken Basin (Shevchenko et al. 2007). According to different estimates, the diameter of this structure is comparable to the diameter of the Moon itself (2500–3500 km). Who, in the history of science and engineering, could be a character of sufficient significance to be commemorated by the naming of this structure remains

Fig. 5.7 Chief engineer of the Special Engineering Bureau #1 (OKB-1), Academician Sergei Pavlovich Korolev (1906–1966)

Table 5.2 The summary of the number of named and unnamed craters of different diameters

Диаметр кратер а, км	Число названных кратер ов	Число н ен азванных кратер ов
21–40	296	3750
41–80	548	1550
81–160	308	416
>160	59	74

an open question. Table 5.2 shows the summary of the number of named and unnamed craters of different diameters (Shevchenko et al. 2009).

The data of Table 5.2 show clearly that in the near future, the fulfilment of "Riccioli's rule" (naming large features after more significant figures) will become a problem. At the same time, it is understood that from all the personalities of science and engineering, who are worthy of memorialisation on the Moon, there are more prominent figures who enjoy particular international fame and recognition, as well as pioneering researchers of the Moon. Over the last three decades, this ethical problem has been decided in different ways.

New names were given to features on the farside of the Moon according to the established tradition. The toponymy of the visible hemisphere remained unchanged. When, at the 17th General Assembly of the IAU (1979), the question arose of the commemoration of the first researcher of the farside of the Moon, Yurii Naumovich Lipsky, it was decided to name a crater in the very centre of the distant hemisphere in his honour (Fig. 5.8). The diameter of this crater is 80 km, and its centre is at the selenographic coordinates of 1.97°S 179.56°W.

Earlier was the first violation of the "untouchability" of the toponymy of the visible hemisphere. On one edition on separate sheets of a large-scale map of Mare Imbrium, a small crater of diameter less than 8 km was named after the famous

Fig. 5.8 Lipsky crater, at the very centre of the far side of the Moon

Fig. 5.9 Famous soviet geochemist and cosmochemist A. P. Vinogradov

physicist L. A. Artsimovich. In 1973, this name was confirmed by the IAU at its 15th General Assembly.

Several years later, a group of hills, an elevated region about 29 km across, was named after the famous soviet geochemist and cosmochemist A. P. Vinogradov (Fig. 5.9). This mountainous structure in the south-west of Mare Imbrium is also on the visible side of the Moon. The coordinates of the center of the formation are 22.35°N 32.52°W. At one time, Academician Vinogradov led the committee on the toponymy of planets of the solar system of the Academy of Sciences of the USSR. Thus, by decision of the 17th General Assembly of the IAU (1979), Mons Vinogradov appeared on the near-side of the Moon (Fig. 5.10).

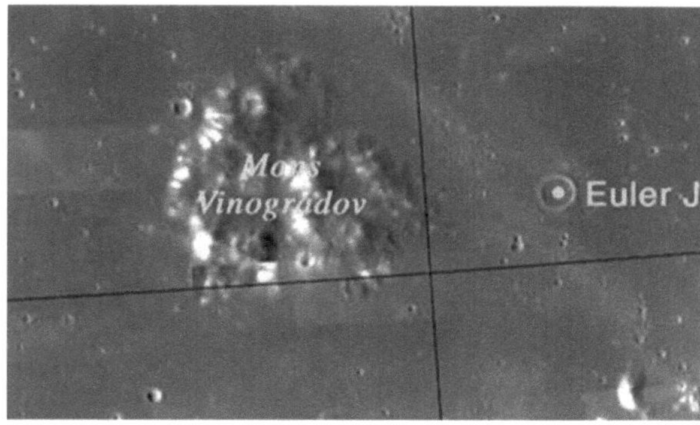

Fig. 5.10 Mons Vinogradov on the near side of the Moon

Fig. 5.11 Mstislav Vsevolodovich Keldysh (1911–1978) (*Source*: www.keldysh.ru/memory/keldysh/photoes.htm)

At the following 18th General Assembly in 1982, it was decided to commemorate the distinguished scientist in the field of space research, who was known for a long time in our country as the chief cosmonautics theorist, Mstislav Vsevolodovich Keldysh (Fig. 5.11). Because of his outstanding contribution to world science, the IAU confirmed the proposal of the Lunar Task Group to name a 32 km crater on the visible hemisphere close to Hercules and Atlas Craters after Keldysh (Fig. 5.12).

It is worth remembering one more unique and relatively recent case when a crater was named on the near-side hemisphere of the Moon. This was the commemoration of the prominent scientist and engineer in the field of rocket construction, Chief Engineer of Rocket Engines and General Engineer of Space Systems, Valentin Petrovich Glushko (1908–1989). The engines created under Glushko's leadership lifted the first and subsequent artificial satellites of the Earth into orbit, the spacecraft of Yurii Gagarin and other cosmonauts, as well as powering the flights of automated spacecraft to the Moon and other planets of the Solar System. Glushko directed the development of the unique reusable spacecraft *Energiya-Buran*, the *Salyut* orbital station, and the base module of the *Mir* space station (Fig. 5.13).

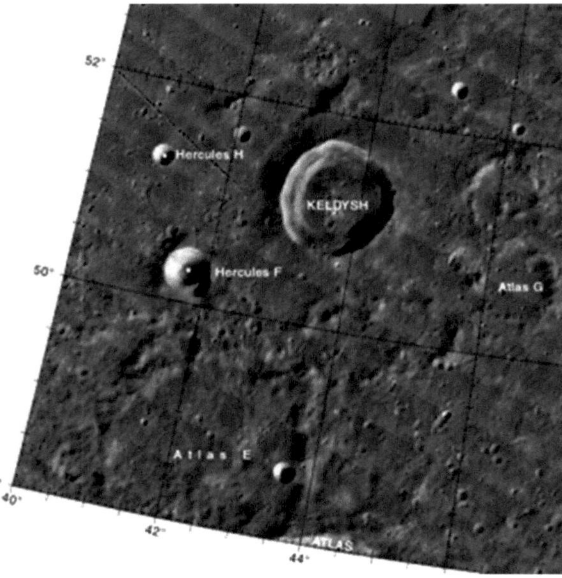

Fig. 5.12 Fragment of a map at scale 1:1,000,000, compiled from images taken by *Lunar Orbiter*. Keldysh Crater is at the centre of the image (*Source*: http://planetarynames.wr.usgs.gov)

Fig. 5.13 Valentin Petrovich Glushko with cosmonauts Yurii Alekseyevich Gagarin and Pavel Romanovich Popovich in his working room. On the Chief Engineer's table is the lunar globe produced by the Sternberg Astronomical Institute of Moscow State University (1968)

The PD-180 engine, notably, which is currently used on the first stage of the US *Atlas III* and *Atlas V* rockets, was based on the first stage engine of *Energiya* developed under the guidance of Academician Glushko at NPO Energomash. The launches executed using the PD-180 engine include the *New Horizons* mission to Pluto (2006), the *Lunar Reconnaissance Orbiter* (2009), the *Solar Dynamics Observatory* (2010), the *Juno* mission to Jupiter (2011), the *Mars Reconnaissance*

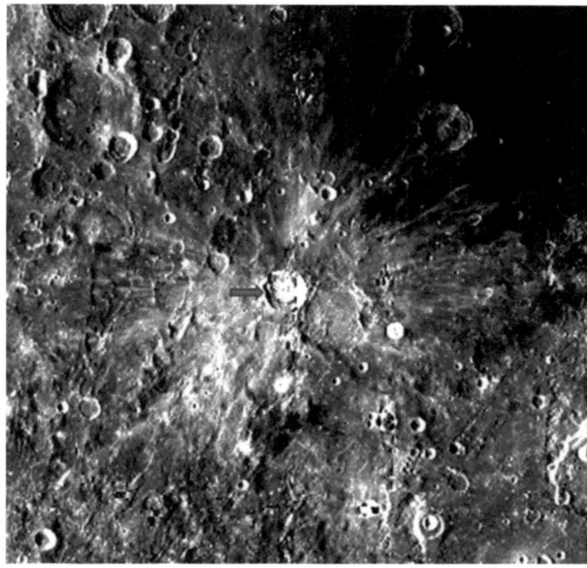

Fig. 5.14 Glushko crater, at the centre of an extended ray system, is located close to the western edge of Oceanus Procellarum (shown with an *arrow*). The image was taken by *Lunar Reconnaissance Orbiter* (*Source*: http://wms.lroc.asu.edu/lroc_browse)

Orbiter (2005), the *Mars Science Laboratory* (2011) and the mission to study Mars' atmosphere, *MAVEN* (2013).

Along with Glushko's well-known achievements in the field of applied cosmonautics, he also made a significant contribution to science: his work of many years on the compilation of thermal constants and the thermodynamic and thermophysical properties of different materials is highly valued by the international community.

In choosing the crater that would be a memorial to Glushko, we made use of several symbolic elements. A crater of 43 km diameter on the lunar near-side, which is at the centre of an extended bright ray system, was proposed to the expert committee. This structure looks like the fiery jets streaming out from the powerful rocket engine designed by Glushko, which launched nearly all of the Soviet and Russian rockets into space, opening a new epoch in the history of humanity for the whole world. In 1994, the proposal of the Lunar Task Group for Nomenclature was approved by a decision by the 22nd IAU General Assembly, and Glushko's name was given to the crater on the nearside of the Moon. Because of its bright rays, extending almost 1000 km across the lunar surface, this crater can be seen at full moon using normal binoculars, not to mention the small telescopes used by many amateur astronomers (Fig. 5.14). Figure 5.15 is a fragment of a map showing the vicinity of Glushko Crater.

Among the unusual decisions in the field of lunar toponomy was the naming of a lunar crater after the famous American geologist and planetary scientist, Eugene Shoemaker (Fig. 5.16). At the time of preparation of the *Apollo* programme, Shoemaker was an astronaut candidate and was preparing for a flight to the Moon. These plans were not fulfilled, but his work in the field of planetology

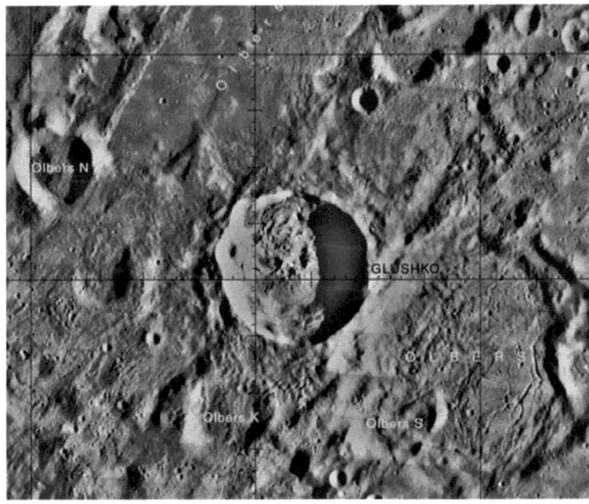

Fig. 5.15 Map fragment showing a detailed image of Glushko crater

Fig. 5.16 Eugene and Carolyn Shoemaker at a symposium in Versailles, France, 1996 (photo: V. V. Shevchenko)

made him widely known. One direction of his work was the study of the phenomenon of cometary and asteroidal impacts onto the Earth and other planets of the Solar System. He was among those who predicted the possible existence of ice deposits in the permanent shadows of polar craters on the Moon. At the same time, he and his wife, Carolyn, were celebrated discoverers of comets. The most famous of these, not only among professionals but also the wider public, was Comet

Shoemaker-Levy, which in 1994 impacted Jupiter having broken up into 20 or so fragments. The theatrical spectacle of the falling and explosion of the cometary nucleus fragments was observed by astronomers and enthusiasts across the world. Numerous works by Shoemaker were dedicated to the study of astroblemes—impact craters on Earth originating from space. During a research expedition to an astrobleme in Australia in the summer of 1997 tragedy struck. As a result of a car crash, Eugene Shoemaker died and his wife was severely injured.

For his services to humanity in space exploration and his study of the influence of extraterrestrial bodies on our planet, and by the wish of his colleagues and family, it was decided to place the ashes of Eugene Shoemaker on the Moon. With the help of NASA specialists, a hermetic capsule was prepared to contain the scientist's ashes. The capsule was placed on board the *Lunar Prospector* probe, which entered lunar orbit in January 1998. At the end of its scientific programme, the project directors took the decision to target the spacecraft at the lunar surface into one of the permanently shadowed polar craters. It was hoped that on impact a gas–dust cloud would form in which Earth-based observers would be able to discover the spectral signature of water (or more specifically the OH line), which would demonstrate the existence of polar ices on the Moon. The cloud was not seen, but the capsule with Shoemaker's ashes reached the Moon on the 31st of July 1999 (Fig. 5.17). When it came to naming a crater after Shoemaker, the natural choice was the one where his ashes lay (Eugene Merle Shoemaker, 1928–1997).

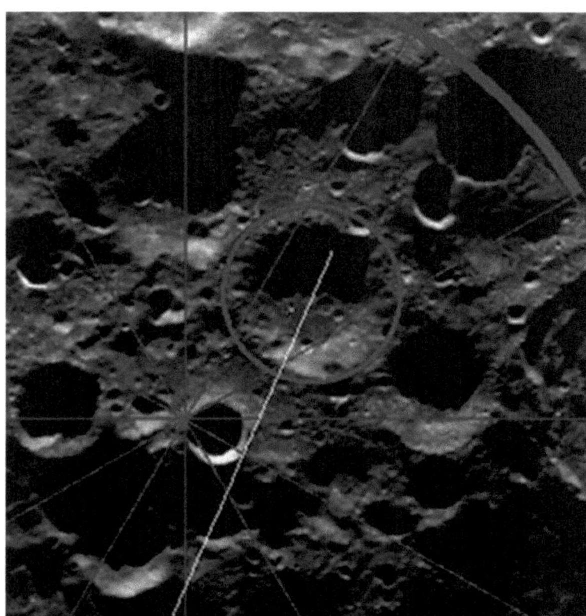

Fig. 5.17 Place of impact of *Lunar Prospector* (projection of impact trajectory shown with a *white line*). Radar image obtained with one of the largest radio telescopes at Goldstone, US (Shevchenko 2001)

On normal images, Shoemaker Crater is not visible because it lies in permanent shadow. It is worth saying a few words here about the choice of the impact site for *Lunar Prospector*. Two years before the event, the *Luna-Glob* project was being developed in Russia. One of the experiments for the spacecraft was designed with the intention of landing on the proposed ice deposits at the lunar South Pole. A successful outcome—direct contact with this mysterious material—would be of epochal significance in the study of the Solar System. Unfortunately, the project was not realised because a policy decision was made to give preference to a less ambitious mission. Interestingly, though, the location chosen for the landing module by astronomers at the Sternberg Astronomical Institute was the very same unnamed crater and cold-trap that—that American specialists chose for the impact of their spacecraft two years later (Galimov 2010).

Thus, an independent analysis by both Russian and international specialists came to the same conclusion: the most probable place to find lunar ice deposits at the South Pole was the cold trap in the permanently shadowed 60 km crater at 88°S 45°E, now called Shoemaker Crater. Figure 5.18 shows a map of the neutron albedo of the South Pole region of the Moon, processed from the results of the Russian instrument *LEND* (Lunar Exploration Neutron Detector) aboard *Lunar Reconnaissance Orbiter*. These data show a significant content of hydrogen in the surface layer on the floor of Shoemaker Crater, which likely indicates the presence of water ice deposits (Shevchenko et al. 2011).

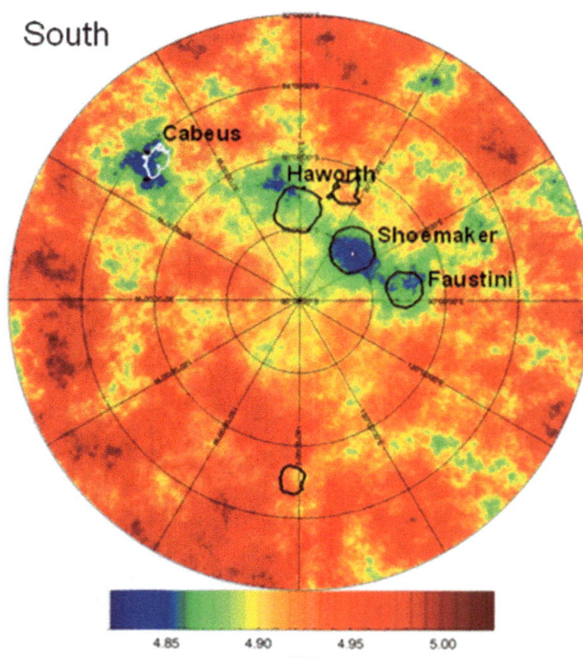

Fig. 5.18 Map of the neutron albedo of the lunar South Pole obtained from measurements of epithermal neutrons by the *LEND* instrument aboard *Lunar Reconnaissance Orbiter*. The map shows the weak flux of epithermal neutrons from Shoemaker crater (Mitrofanov et al. 2010, 2011)

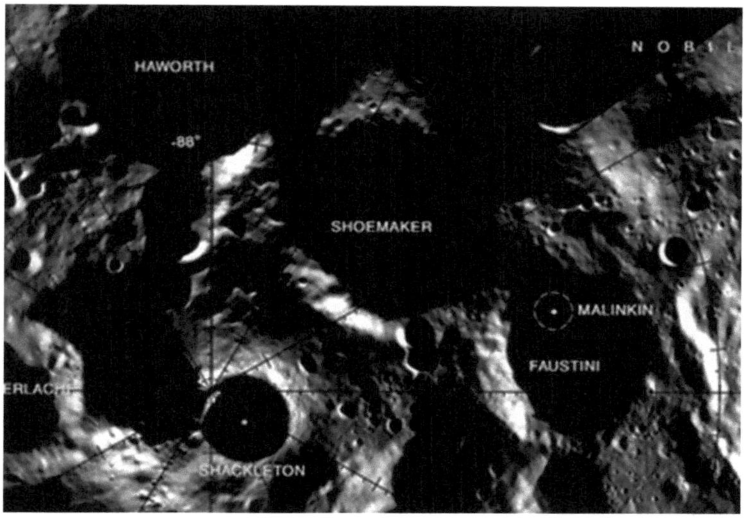

Fig. 5.19 Fragment of a map of the lunar South Pole region including Shoemaker crater (http://planetarynames.wr.usgs.gov)

By a decision of the Working Group for Planetary System Nomenclature, taken at the 26th IAU General Assembly (Prague, 2006), it was stated in particular that "*Specific names may be proposed by the Task Groups or the Working Group. Suggested names may be accepted from the scientific community or the general public. However, the application of names to specific features must adhere to the appropriate category and be consistent with the Working Group's required goal to maintain cultural, national, and geographical diversity, and to name features only when they have special scientific interest*".

The appearance of Shoemaker Crater on lunar maps is a good example of precise adherence to this rule. The name not only fully satisfies the requirements with respect to the person remembered but also specifies a lunar structure of high scientific interest (Fig. 5.19).

In Memory of Tragedies in Space

The new times have left traces on the Moon, not only of successes and glory but also of the inescapable losses which accompany every penetration into the unknown. The tragic event at the time of training in the low-pressure chamber of the Institute of Aviation and Space Medicine in the programme of the Cosmonaut Training Center (today called the Gagarin Research and Test Cosmonaut Training Center) led to the death of the youngest trainee from the first ever cosmonaut selection (Fig. 5.20).

Fig. 5.20 Trainee cosmonaut Valentin Vasiliyevich Bondarenko (1937–1961) (http://www.astrolab.ru/cgi-bin/img.cgi?i=1466.jpg)

The tragedy occurred only 19 days before Yurii Gagarin's historic flight on the 23rd March 1961. For a long time, this event was kept in strict secrecy. Only at the beginning of the 80s, did any information about the tragic death of Bondarenko appear in print. Later, in 1986, the famous journalist Yaroslav Golovanov wrote about it. One of the most important American lunar planetary researchers, Hal Mazursky, who was then chairman of the IAU Working Group on Planetary System Nomenclature, addressed the chairman of the Lunar Task Group, Vladyslav Shevchenko, with the proposal to name a crater on the lunar farside after Bondarenko.

The administration of the Cosmonaut Training Center showed understandable caution in providing an answer to the request for the required biographical information about Bondarenko. An appeal to Academician Glushko, who as General Engineer supported Shevchenko's request, proved decisive and the information was given. The 21st IAU General Assembly held in 1991 confirmed the name Bondarenko for a 28 km diameter crater on the farside of the Moon close to Tsiolkovsky Crater (Valentin Vasilyevich Bondarenko, Soviet trainee cosmonaut, 1937–1961). Figure 5.21 shows a map fragment of the region of Bondarenko Crater.

Unfortunately, a few years later a similar tragedy befell a crew of American astronauts while preparing for the first manned flight of the *Apollo* programme on 21st February 1967. A fire broke out in the command module and all the crew were killed. The accident occurred on 27th January 1967 during ground tests at launch complex No. 34 of the Kennedy Space Center. The astronauts were Edward White, Virgil "Gus" Grissom and Roger Chaffee (Fig. 5.22). *Apollo-1* was the name later given to the mission which did not take place (AS-204). While training in the hermetically sealed module filled with a high oxygen content atmosphere, a chance spark ignited the fire. The astronauts could not be saved (Fig. 5.23). The first manned flight, *Apollo-7*, took place only in October 1968 after an investigation and the introduction of many improvements to the design of the lunar spacecraft.

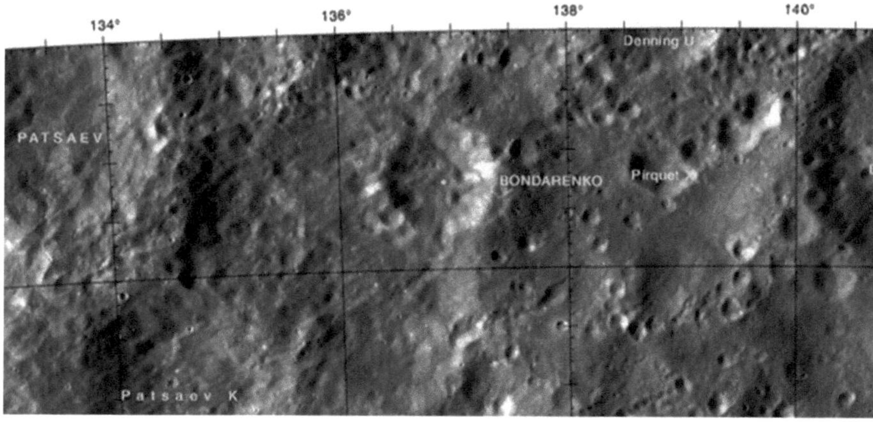

Fig. 5.21 Fragment of the lunar map of the region of Tsiolkovsky crater. Bondarenko crater is located at 17.24°S 136.89°E (http://planetarynames.wr.usgs.gov)

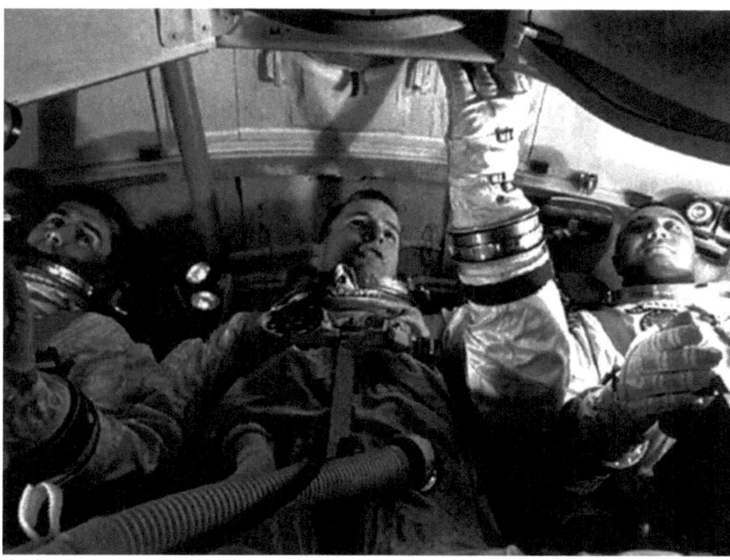

Fig. 5.22 Astronauts Edward White, Virgil "Gus" Grissom and Roger Chaffee in the cockpit of *Apollo-1* (http://www.space.com/10674-apollo-1-fire-nasa-disaster.html)

A decision of the 24th IAU General Assembly in 1970 named three craters in the vicinity of the Apollo basin on the lunar farside in memory of each of the astronauts (Fig. 5.24).

Sadly, these losses among the lunar pioneers were not to be the last. In the same year, 1967, Vladimir Mikhailovich Komarov was killed when, after 19 orbits of the Earth, the *Soyuz-1* parachute failed to deploy during descent. Georgii Dobrovolsky, Vladislav Volkov and Viktor Patsayev died in 1971 when their *Soyuz-11* spacecraft

Fig. 5.23 *Apollo-1* cockpit after the fire of 27th January 1967 (http://www.space.com/10674-apollo-1-fire-nasa-disaster.html)

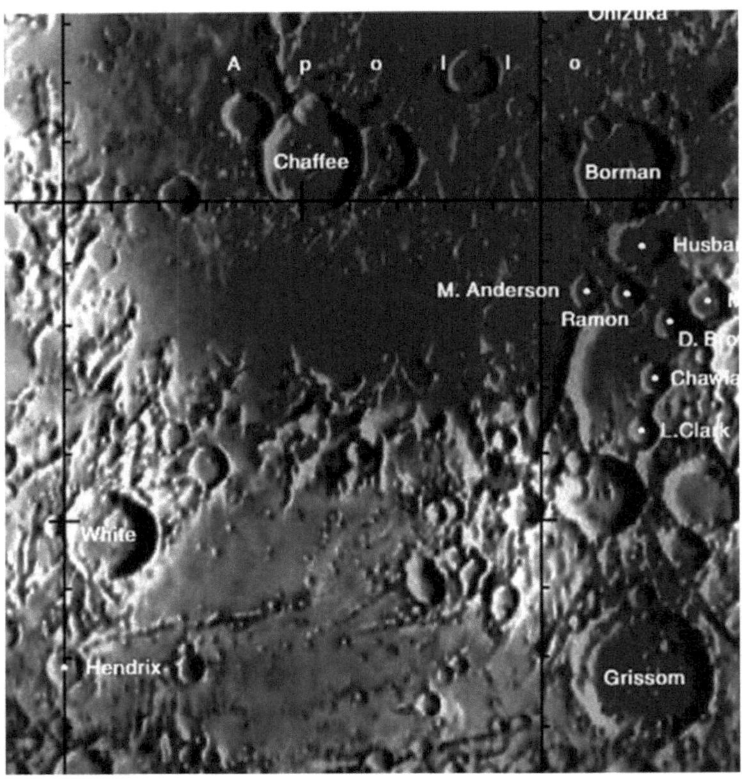

Fig. 5.24 Fragment of 1:10 million scale shaded relief and colour-coded topography map showing the Apollo basin and the craters named after Grissom, White and Chaffee (Image courtesy of the USGS Astrogeology Science Center)

depressurised during re-entry preparation. Their mission had been the first successful docking with a space station, *Salyut-1*, where they had spent 22 days. In 1986 and in 2003, as a result of failures of the reusable space shuttles, *Challenger* and *Columbia*, fourteen crew members did not return to Earth. The IAU has named lunar craters in memory of all these lost spacefarers.

Happy Exceptions to the Rule

The first substantial deviation from the rules adopted by the IAU was the naming of the largest mare feature on the lunar farside. As mentioned earlier, lunar mare are normally named after weather phenomena or emotional states. Thus, we have Oceanus Procellarum (Ocean of Storms) or Mare Imbrium (Sea of Rains), Mare Tranquillitatis (Sea of Tranquility) or Mare Crisium (Sea of Crises). In confirming the first 18 names of features on the lunar farside, the 11th IAU General Assembly held in 1961 in Berkeley acceded to the wishes of the discoverers and, in spite of the convention, accepted the name Mare Moscoviense (Sea of Moscow) for the newly discovered mare feature. This name has firmly entered the lunar nomenclature.

Another exception from the established conventions was the naming of a series of craters on the farside in honour of living cosmonauts and astronauts. Academician Glushko, who took an active role in the creation of lunar and planetary toponymy, proposed at the end of the 60s to name a series of lunar craters after space explorers who participated in flights of particular significance. His reasoning was that the achievements of these men should not be re-evaluated in a subsequent time. The astronomical community supported this proposal, and the 14th IAU General Assembly (1970) confirmed the new names of this character. In the vicinity of Mare Moscoviense appeared craters named after Titov (first 24 hour space flight), Tereshkova (first woman in space), Belyaev and Leonov (first spacewalk) and others (Fig. 5.25).

The IAU rules strictly require an international approach to the choice of names for lunar features, so an equivalent series of names was proposed reflecting the pioneering achievements of American astronauts. The basin structure of diameter 524 km located on the farside with centre coordinates 35.7°S 151.5°W was named Apollo in honour of the spacecraft which brought the first human to the Moon (Fig. 5.24). Around this basin there are three craters which were named Borman, Anders and Lovell after the crew of *Apollo-8*, the first piloted orbit of the Moon in 1968 (Fig. 5.26).

It goes without saying that the lunar toponymy could not but reflect such an important event in the history of humanity as the first human visits to the Moon. On the nearside, close to the landing site of the manned spacecraft *Apollo-11*, three craters were named by a decision of the 14th IAU General Assembly in 1970 after the first men to step on the Moon, Neil Armstrong and Edwin "Buzz" Aldrin, and the commander of the orbiting module, Michael Collins. Figure 5.27 shows the legendary crew of *Apollo-11* photographed at the Kennedy Space Center in Florida,

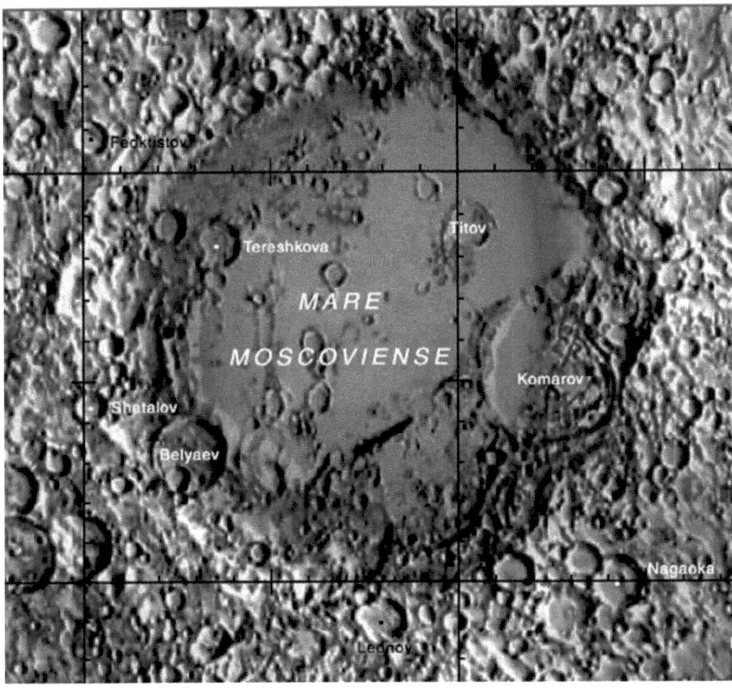

Fig. 5.25 Fragment of lunar map showing the region of Mare Moscoviense (http://planetarynames.wr.usgs.gov)

US, during preparations for the space expedition. Figure 5.28 shows craters named after Armstrong, Aldrin and Collins.

At the time of the 12th Committee on Space Research (COSPAR) General Assembly, held in Leningrad (present-day St. Petersburg) in 1970, Neil Armstrong also visited Moscow and the Lomonosov Moscow State University. There he met those who compiled the lunar maps and globes and signed his autograph on one globe on Mare Tranquillitatis, where the craters named after Armstrong, Collins and Aldrin are found (Fig. 5.29).

Difficult Decisions

The rules adopted by the IAU purposely exclude any possible influence of politics in the process of deciding names for lunar features, as well as for other bodies of the Solar System. Names of politicians, military leaders, contemporary philosophers and religious figures are not accepted for consideration. But in a string of cases, political aspects were unavoidably mixed into the process of creating the lunar toponymy.

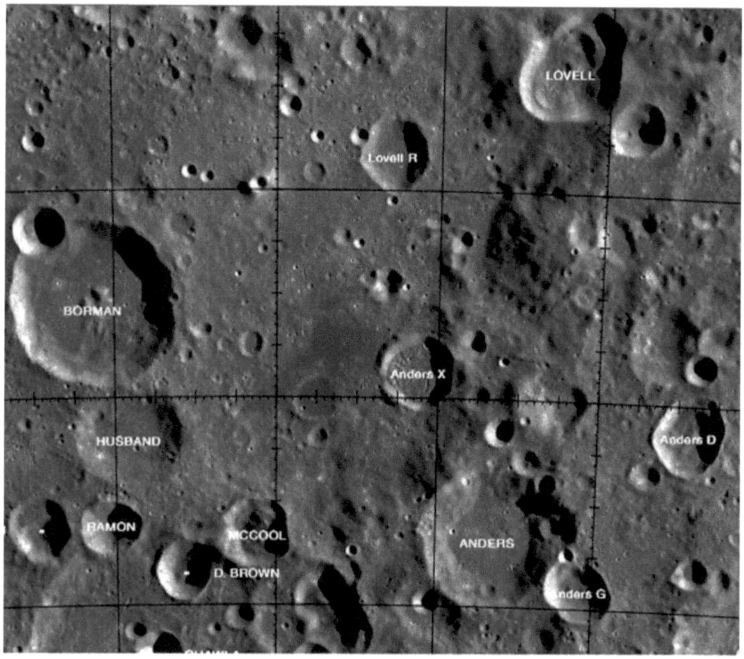

Fig. 5.26 Fragment of lunar map showing the craters named after the crew of *Apollo-8* (http://planetarynames.wr.usgs.gov)

Fig. 5.27 From *left* to *right*: Neil Armstrong, Michael Collins and Edwin Aldrin in Florida at the *Saturn-5—Apollo-11* launch complex (Photo NASA)

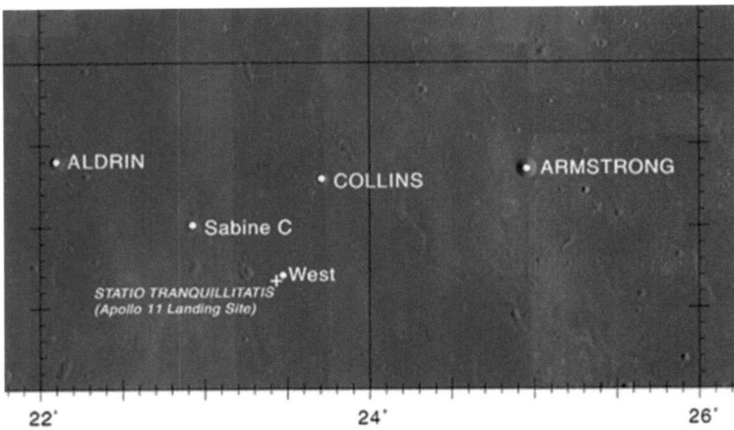

Fig. 5.28 Map fragment showing Armstrong, Aldrin and Collins craters (http://planetarynames.wr.usgs.gov)

Fig. 5.29 Neil Armstrong signs his autograph on a globe of the Moon, which was produced at the Sternberg Astronomical Institute of the Lomonosov Moscow State University (Photo: V. V. Shevchenko, 1970)

The talented Russian scientist, Yurii Vasilievich Kondratyuk, independently of Tsiolkovskii, derived the fundamental equation of rocket motion, made a design for a four-stage rocket powered by oxygen and hydrogen fuel and proposed the use of the gravitational fields of planetary bodies for the acceleration and braking of spacecraft for flights through the Solar System, among other things (Fig. 5.30). The most famous of these was the so-called *Kondratyuk-trajectory*. The idea was to fly to the Moon using a single launch rocket with the spacecraft dividing in lunar orbit into a landing and take-off module and an Earth return module. This scheme for a manned flight to the Moon was proposed in his earliest work, written in 1916–1919 and published by the Institute for the History of Science and Technology of

Fig. 5.30 Yurii Vasilievich Kondratyuk (1897–1942?) (http://kraeved.ngonb.ru/node/3427)

Fig. 5.31 John Houbolt explains his proposed scheme for the *Apollo* mission. The scheme repeats that of Yurii Vasilievich Kondratyuk in his early work *To those who will read in order to build*. This work was translated into English in 1965 (Photo: NASA, 1962)

the USSR Academy of Sciences only in 1964. The idea was realised by the *Apollo* project many years later.

In a series of publications, it was stated that the *Kondratyuk-trajectory* was directly adopted from the English translation of his works and proposed for practical realisation by John Cornelius Houbolt, a member of the *Apollo* project (Sheridan 1969). Evidently, this is not how it happened and Houbolt came to the same idea independently, which nevertheless does not diminish the significance of Kondratyuk's engineering foresight (Fig. 5.31).

Kondratyuk proposed "it is more efficient not to take the whole spacecraft onto the planet, but to leave it as a satellite (around the planet), and to visit the surface with only the part which is necessary for landing and for the return trip to the orbiting spacecraft". Using just a small part of the rocket for landing significantly reduces the required fuel and, consequently, the launch mass of the original rocket. Houboldt's calculations demonstrated this, and the scheme was adopted for the *Apollo* project.

Kondratyuk's name, as a soviet pioneer of cosmonautics, was proposed for commemoration on the Moon in the 1960s. For a long time, however, the initiative received no official support. It turned out that the scientist, in order to escape responsibility for a short period serving in the White Army, had been living under the name A. I. Shargei since 1921, making use of the identification documents of a family friend who had died. In the 30s, once more living by the name of Kondratyuk, he was subjected to repression as a traitor. At the start of the war in 1941, he volunteered for the front line. After the first battles, there is no more record of him. According to one source he was killed in October 1941; another suggests he went missing in action at the beginning of 1942. This latter formulation in the legal terminology of the time meant that he "surrendered into captivity", i.e., he became a prisoner. This report cast a dark shadow over Kondratyuk's official biography for many years.

Only in 1970, the Judicial Board on criminal cases of the RSFSR Supreme Soviet by declaration No. OS-70-8 on 26th of March rehabilitated Kondratyuk for lack of evidence of a crime. In the same year, his name was presented to the IAU and by a decision of 14th General Assembly, it was given to a crater on the lunar farside to the west of Tsiolkovsky Crater (Fig. 5.32).

The historical role of the famous rocket engineer, Werner von Braun, who began his career in Germany in the 1930s and continued his work in the United States, is ambiguous. After his death, when the question arose of commemorating him on the Moon, opponents of the proposal recalled the *V2* rockets he had designed, which were fired at London during the Second World War, and his collaboration with the Nazis at the beginning of his career building rockets (Fig. 5.33).

Strong objections to the idea were raised by the prominent Canadian astronomer, Peter Mackenzie Millman (1906–1990), who by that time had become the first chairman of the IAU Working Group for Planetary System Nomenclature. Having begun his scientific career in 1933 at Dunlap Observatory of the University of Toronto, Millman joined the Royal Canadian Air Force at the outbreak of the World War II and fought throughout the war. He only returned to scientific work in 1946 (Fig. 5.34).

In his objections, Millman emphasised that, as a veteran of World War II and having fought against fascism, he could not on moral grounds agree to the commemoration of Werner von Braun on the Moon. Among other voices speaking against the proposal was that of the first researcher of the lunar farside mentioned previously, professor of Lomonosov Moscow State University, Yurii Naumovich Lipsky. Like Millman, Lipsky was a war veteran (Fig. 5.35), having fought at battles on the Voronezh and Ukrainian fronts, in Poland, Czechoslovakia and Germany, and been wounded three times. He returned to scientific work in September 1945.

During the discussions of this problem, Lipsky was a member of the Lunar Task Group. Supporters of the proposal to name a lunar crater after von Braun pointed

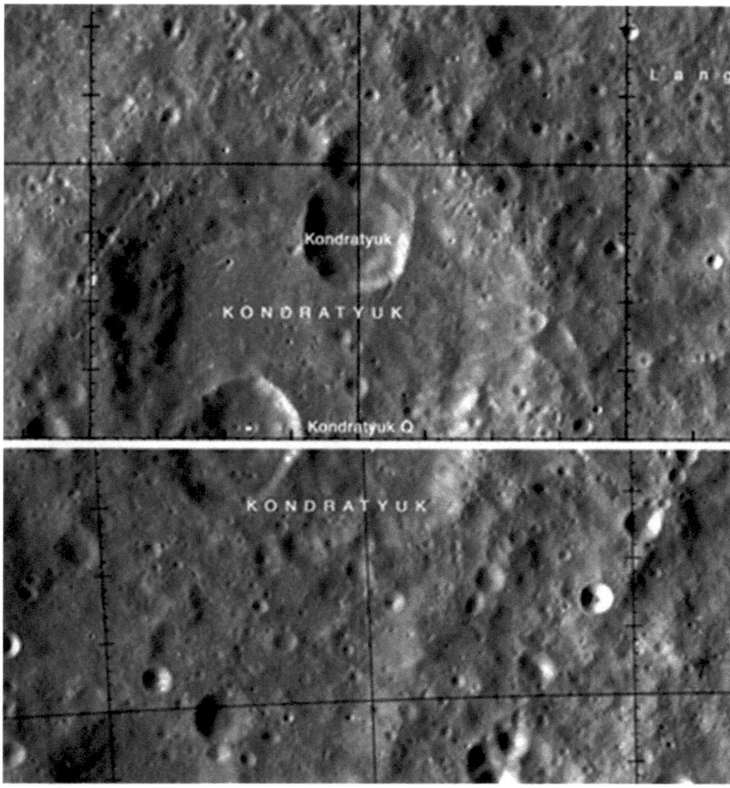

Fig. 5.32 Mosaic of two map sheets of the lunar farside in the region of Kondratyuk Crater. The crater is 98 km in diameter with its centre at 15.33°S 115.8°E (http://planetarynames.wr.usgs.gov)

out that the majority of the creative episode of the engineer's life had been dedicated to modern rocket technology and, in particular, the *Apollo* project. It was von Braun's labours which laid the path for humanity to reach the Moon (Fig. 5.36). The engineer repeatedly explained the first steps of his career in Germany by the insurmountable external circumstances and his immature age. In particular, he noted that for expressing the idea to use rockets for space exploration instead of military application, he had been arrested by the Gestapo in March 1944 and been imprisoned for some time.

The discussions continued for seventeen years until the IAU, at its 22nd General Assembly in 1994, took the decision to name a lunar crater after von Braun (Fig. 5.37).

As mentioned before, the IAU rules do not permit a feature name to be changed, other than names made by the addition of a Latin letter, which are considered temporary. In all other cases, the name is permanent, as the names retained from Riccioli's time demonstrate. However, in the recent history of lunar toponymy, there was a case when, on moral grounds, this rule had to be broken.

Fig. 5.33 Werner von Braun (in civilian clothes) among Nazi officers, 1942 (Photo: NASA)

According to a decision of the 16th IAU General Assembly held in 1976 in Grenoble, France, a small crater of diameter 5.78 km in Mare Cognitum received a new name. The crater Euclides D in the new edition of the lunar toponymy was

Fig. 5.34 Peter Mackenzie Millman, chairman of the IAU Working Group for Planetary System Nomenclature from 1973 to 1982 (in the centre background can be seen the lunar globe produced at the Sternberg Astronomical Institute)

Fig. 5.35 Professor of Lomonosov Moscow State University, Yurii Naumovich Lipsky, under whose leadership several editions of lunar maps and globes were produced at the Sternberg Astronomical Institute

called Eppinger Crater in honour of the Austrian-Czechoslovakian doctor, Hans Eppinger (1878–1946), who became famous for his medical research into the treatment of kidneys. Eppinger was a professor at the Universities of Freiburg, Cologne and Vienna and was invited for some time to the USSR to treat Stalin. After the new version of the lunar toponymy and the maps on which it was reflected reached a wider audience, the IAU began to receive enquiries about Eppinger. It came to light that Dr. Eppinger, during World War II, had taken part in experiments on prisoners in the Nazi concentration camp at Dachau. In 1946, Eppinger was summoned for trial at Nuremberg. However, a month before the planned hearing, he killed himself.

After verifying the information in 2002, the Lunar Task Group—going against the IAU rules but considering the special circumstance—proposed removing the name from the lunar nomenclature and lunar maps. The IAU WGPSN supported the decision, and finally the 27th General Assembly in 2009 took the final decision to revert the crater's name to its former Euclides D. The single known mistake in the 300-year history of lunar toponymy had been corrected.

Fig. 5.36 Werner von Braun (1912–1977) beside the rocket engines of the first stage of a *Saturn V* (http://www.apolloarchive.com/apollo_gallery.html)

Contemporary Status and Perspectives

For many years, the procedure for naming lunar features has involved the compilation of a preliminary database of names. At the present time, the database contains more than 250 names of prominent scientists and engineers, including 45 Nobel laureates. Figure 5.38 shows the distribution of names in the database according to

Fig. 5.37 Von Braun Crater on the western edge of Oceanus Procellarum, 64 km in diameter with centre coordinates 41.04°S 78.08°W (http://planetarynames.wr.usgs.gov)

the scientific field. This information illustrates the cultural and historical significance of the lunar toponymy. The most prominent achievements of our time frequently occur in biology and physics, which is reflected in the diagram. Space science is still a young field of knowledge, which has not yet attained a sustained level of development.

As mentioned above, the current expansion of the lunar toponymy has encountered many objective difficulties connected with the structure of the lunar relief (Fig. 5.38).

The application of "Riccioli's rules" is limited to the number of unnamed large features. Figure 5.39 shows the breakdown according to diameter of named and yet unnamed lunar craters. It follows from the diagram that the main possibility for extending the lunar toponymy relates to smaller sized craters. It is worth noting that, in the diameter range of 10–20 km, around 7060 craters are named with letters, which are adopted by the IAU as a temporary nomenclature.

The lunar toponomy is naturally extended by the naming of features of particular scientific interest. The process is driven, of course, by new findings in research of the nature of the Moon.

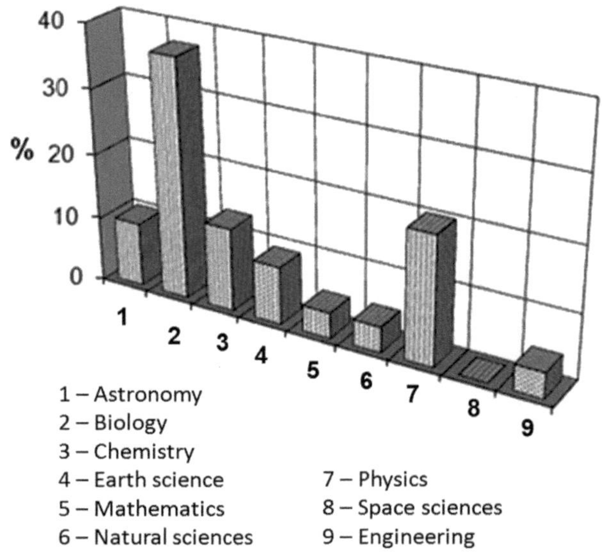

Fig. 5.38 Distribution of names of prominent scientists and engineers included in the lunar database, categorised by scientific field

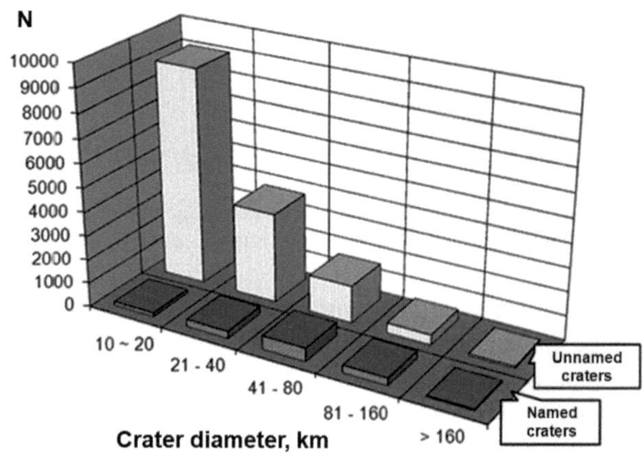

Fig. 5.39 Distribution by size of named and unnamed lunar craters

The discovery of the compositional signature of cometary material (Colaprete et al. 2010) in the low temperature polar deposits once again drew attention to the diffuse structures known as "lunar swirls", which might be the trace of cometary impacts into the Moon.

Consequently, a systematisation of the information about these anomalous structures requires giving them names, identifying their locations and nature. At present, only one such structure on the near-side in Oceanus Procellarum carries a name: Reiner-γ. The name was given in relation to the nearest crater, Reiner, with the addition of a greek letter (Fig. 5.40).

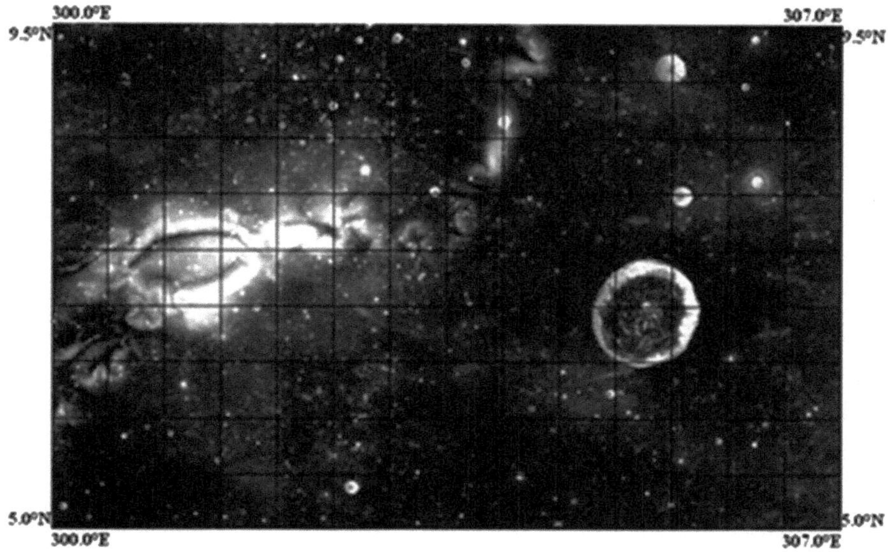

Fig. 5.40 The diffuse structure ("lunar swirl") Reiner-γ (*left*) and Reiner Crater (*right*) imaged by *Clementine*

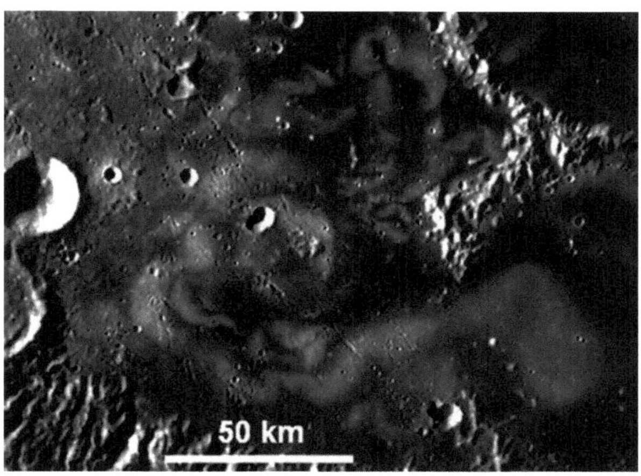

Fig. 5.41 Complex of diffuse structures in Mare Ingenii on the lunar farside (Photo: NASA)

There are swirls which are more substantial in size in a variety of forms and shapes on the farside of the Moon. Figure 5.41 shows an image of a complex of the diffuse structures seen in a region of Mare Ingenii.

The problem of naming these structures remains difficult. The lunar swirl Reiner-γ has a well-defined position and a compact structure. The numerous separate fragments of diffuse structures on the farside, often merging one into another as is seen in Fig. 5.41, provides a challenge for those who would compile a toponymy of these features.

Conclusion

The process of creation of the lunar toponymy has passed through several stages. The early period is distinguished by a chaos of approaches and the complete arbitrariness of the map compilers. Then the international community of astronomers tried to introduce a systematic approach, but the distribution of some or other names across the lunar surface was administered by those who compiled the maps and catalogues.

In recent decades, the logic of lunar research has forced those who compiled the lunar toponymy to adopt a strict systematisation in the distribution of names in different regions of the Moon, which is guided not only by the needs of cartographers but by scientific necessity.

In concluding this essay on lunar toponymy, we should underline once more the cultural and historical significance of the system of names on bodies of the Solar System, including the natural satellite of the Earth. The contemporary, well thought-out approach to the construction of the toponymy for the surfaces of the Moon, planets and small bodies ensures completeness and reliability in covering the past and present history of the development of scientific, technical and cultural knowledge. In hundreds of years, our descendants will assess our time, in part, from this source of information.

References

Blagg, M., & Müller, K. (1935). *Named lunar formations* (196 p). London: Percy Lund, Humphries & Co. Ltd.

Colaprete, A., Schultz, P., Heldmann, J., Wooden, D., Shirley, M., Ennico, K., et al. (2010). Detection of water in the LCROSS ejecta plume. *Science, 330*, 463–468.

Galimov, E. M. (2010). *Zamuslu i Proschetu*. Izdatelstvo Editorial YRSS. Moskva (124 s).

Kopal, Z., & Carder, R. W. (1974). *Mapping of the Moon*. Dordrecht/Boston: D. Reidel Publishing Company (240 pp).

Mitrofanov, I. G., Litvak, M. L, Sanin, A. B., Golovin, D. V., Boynton, W. V., Chin, G., et al. (2011). *Neutron suppression regions at lunar poles, as local areas of water-rich permafrost*. LPS-42. # 1787.

Mitrofanov, I. G., Sanin, A. B., Boynton, W. V., Chin, G., Garvin, J. B., Golovin, D., et al. (2010). Hydrogen mapping of the lunar south pole using the LRO neutron detector experiment LEND. *Science, 330*, 483–486.

Rodionova, Zh. F., Karlov, A. A., Skobeleva, T. P., Konotopskya, E. V., Shevchenko, V. V., Kozubsky, K. E., et al. (1987). *Morfologicheskiy Katalog Kraterov Luny*. Pod Obschey redaktsiey Shevchenko V.V. Izdatelstvo Moskovskogo Universiteta. Moskva (173 s).

Sheridan, D. (1969). If there is any idea we must push, it's this one. *Life, 66*(10), 20–24.

Shevchenko, V. V. (2001). Lunar Prospektor Pogib, Problemu ostalic'. *Zemlya I Vselennaya, 1*, s. 23–33.

Shevchenko, V. V. (2009). *Moon research for half a century. Fifty years of space research* (pp. 127–152). Moskva: Physmatlit.

Shevchenko, V. V., Chikmachev, V. I., & Pugacheva, S. G. (2007). Structure of the South Pole–Aitken Lunar Basin. *Solar System Research, 41*(6), 447–462.

Shevchenko, V., El-Baz, F., Gaddis, L., Hiesinger, H., Shkuratov, Yu., Whitaker, E., et al. (2009). The IAU/WGPSN lunar task group and the status of lunar nomenclature, LPS-40, #2016.

Shevchenko, V. V., Mitrofanov, I. G., Kozlova, E. A., Shangaraev, A. A., & The LEND Science Team. (2011, October 10–14). *Epithermal flux depression and PSR in Shoemaker crater*. Abstract No. 2MS3-LS-09b, The Second Solar System Symposium "Moons of Planets", Moscow, Russia.

Chapter 6
Cartography of Mars

Abstract The first flyby of Mars was made by the Soviet *Mars 1*, launched from Baikonur Cosmodrome on 1st November 1962. The first soft landing was made by *Mars 3*. In 1977, TsNIIGAiK published *Map of a Region of Mars* at 1:5,000,000 scale in normal conical conformal Lambert–Gaussian projection based of phototelevision images from *Mars 4* and *Mars 5*. A collective of geologists compiled geomorphologic and tectonic maps of Mars at 1:20,000,000 scale in 1981. They used *Topographic Map of Mars* issued by the USGS in 1976 as the geographical basis. In 1989, a small production run of Mars globes of 34 cm diameter was produced, followed by a 10,000 unit production at 26 cm diameter. A *Hypsometric Map of Mars* at scale 1:26,000,000 was compiled using *Mars Global Surveyor MOLA* altimetry in 2004. The *Hypsometric Globes of Mars* of diameter 21 and 15 cm were compiled using the thermoplastic production process.

The Italian astronomer Giovanni Virginio Schiaparelli (1835–1910) is famous as the discoverer of canals on Mars. Few, however, know that during 1859–1860 he spent a period in Russia at the Pulkovo Observatory studying under the guidance of the director, Otto Vasilyevich Struve (1819–1905), and that in 1874 he was elected a corresponding member of the St. Petersburg Academy of Sciences (Fig. 6.1). Over a period of 45 years, Schiaparelli corresponded with Struve, addressing him as "your excellency" and closing his letters with "your boundlessly loyal friend and son" (e.g., in the letter from 19th February 1879). At the times of the Mars oppositions in 1877 and 1879, he discovered linear features on the surface of Mars, calling them *canali*. In his sketches, Schiaparelli denoted dark and light features with the terms: sea, bay, lake, swamp, lowland, cape, strait, spring and plain and took the names for these from old maps of the Earth or from mythology (Blunck et al. 1993). Some of these names are still used on modern maps. From the published correspondence between Struve and Schiaparelli (Abalakin 2005), it appears that Otto Vasilyevich took a direct role in creating the Martian nomenclature. Since this is of historical importance, we present some extracts from the correspondence:

O. V. Struve – G. V. Schiaparelli
28th November 1877, Pulkovo

> *"...I fully approve your proposal that the astronomical community should become, to a certain extent, the godfather of the nomenclature of clearly discernable marks on Mars, and*

V. Shevchenko et al., *Lunar and Planetary Cartography in Russia*, Astrophysics and Space Science Library 425, DOI 10.1007/978-3-319-21039-1_6

am very much prepared to present this proposal to the Presidium. For this purpose, it would be desirable that we are able to make a definite proposal, and to formulate this there is no-one more knowledgeable and qualified than you. To use simply numbers and letters would be insufficient, since they make less impression on the memory than names. There are already sufficient gods, goddesses and demigods in the heavens, as there are astronomers on the Moon. Would you not like, if you would permit me to propose it, to have Greek philosophers or, since we are talking about Mars, Greek and Trojan heroes from the Iliad? The Greeks, since they lived on ships, could be placed on the seas, and the heroes of Troy on the supposed firm ground..."

(p. 103)

G. V. Schiaparelli – O. V. Struve
4th January 1878, Milan

"...If I send you these names, it is only to show that I have almost followed your idea to take names from the Iliad – you will even find the names of rivers of Ilion. However, I am prepared to accept any system of nomenclature which would be approved by the astronomical community. It would be good if some scholar with a sense of aesthetics would like to bring all of this into proper order..."

(pp. 105–106)

O. V. Struve – G. V. Schiaparelli
12th June 1878, Pulkovo

"...Indeed, what could be more appropriate than the transference to Mars of mythological symbols from the earliest times of geography at the moment when areography is at the very same stage of development as was once the geography of the ancient Greeks? If I should make any correction, I would propose to you the following: taking into account that, with time, you are likely to find and name a significant number of details, you could already give well-established and well-defined features [of the surface of Mars] the names of countries and seas of the Mediterranean, which were known, of course, to the ancient Greeks, and denote the features more distant from the central region of marks, whose outlines you remain guessing to a certain extent, with the mythological names..."

(p. 109)

Fig. 6.1 Photographs of the Italian astronomer Giovanni Schiaparelli and the director of the Pulkovo Observatory, Otto Vasilyevich Struve

Many astronomers could not see the *canali* as, for example, Nathaniel Green, who mapped Mars in 1877, or Nicolas Flammarion's student, Eugène Antoniadi, who later made a very detailed map of Mars. On Percival Lowell's maps, however, there were more than 600 canals (compared to Schiaparelli's 113). Lowell was convinced that another centre of intelligent life in the Solar System would be found on Mars. Schiaparelli considered this idea unfounded. However, when the US spacecraft *Mariner 9* transmitted images of the whole surface of Mars in 1971–1972, it became clear that only a small fraction of the mapped *canali* coincided with valleys, ridges, crater chains and linear albedo features, and that most of them had no relation to either topographic or albedo characteristics of the planet's surface (Sagan and Fox 1975).

The first flyby of Mars was made by *Mars 1*, launched from Baikonur Cosmodrome on 1st November 1962. On 19th June 1963, the spacecraft passed over the surface of the red planet at a distance of 193,000 km. It transmitted a large volume of telemetry and received over 3000 commands from Earth. *Mars 2*, having covered a distance of about 470 million km, went into orbit around Mars on 27th November 1971. This was two weeks after the American spacecraft *Mariner 9* became the planet's first artificial satellite. The descent module of *Mars 2* was destined to become the first man-made object to reach the surface. The first soft landing was made by *Mars 3*. Figure 6.2 shows the region of the landing site

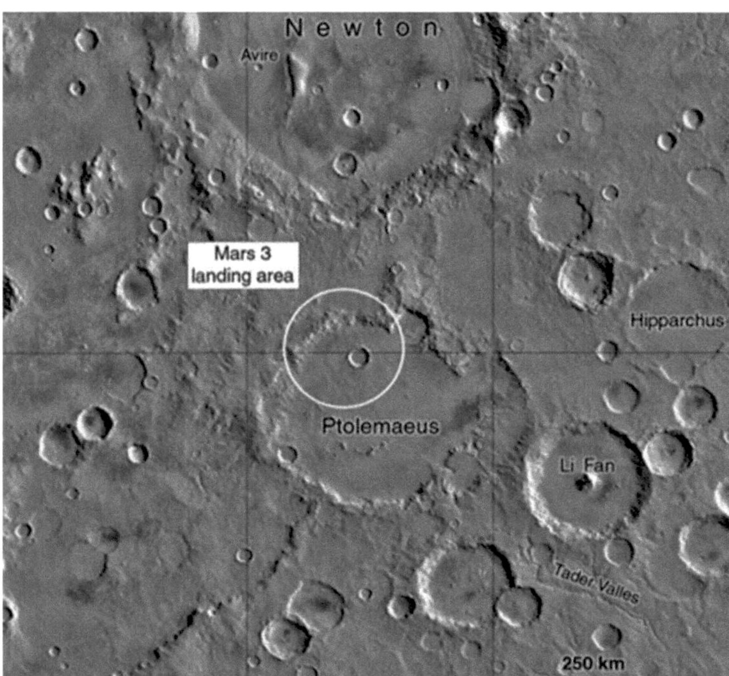

Fig. 6.2 (**a**) Mars 3 landing area, uncertain by 150 km. (**b**) Probable Mars 3 target area on a U. S. Air Force (ACIC) map based on Mariner 4 images. (**c**) Mars Odyssey THEMIS infrared mosaic of the landing area (Stooke 2012a)

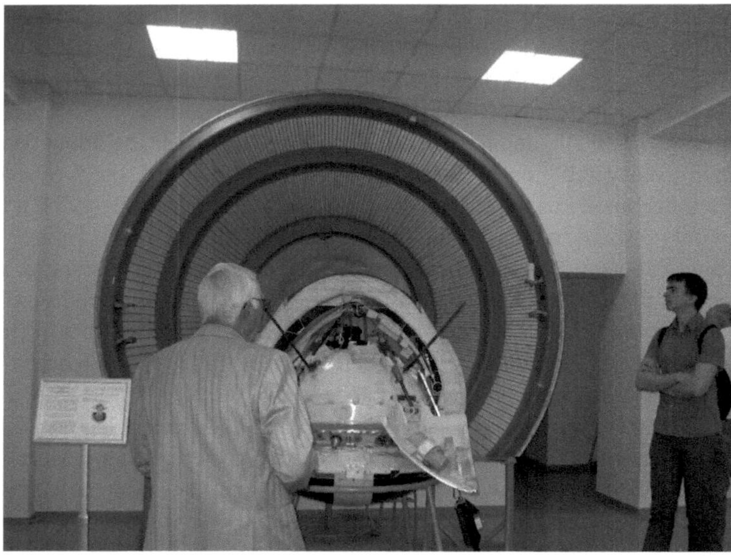

Fig. 6.3 Model of the descent module of *Mars 3* with its heat shield at the Lavochkin Museum, Moscow (Photo: Zh. F. Rodionova)

(Stooke 2012a). It is interesting to note that the apparatus was discovered in images from *Mars Reconnaissance Orbiter* (*MRO*) by Vitalii Egorov from St Petersburg, with the support of A. T. Basilevsky (GEOKhI) and NASA.

A general view of the descent module of *Mars 3* is shown in Fig. 6.3. The apparatus had a huge heat shield which was jettisoned before landing.

In 1977, TsNIIGAiK published *Map of a Region of Mars* at 1:5,000,000 scale in normal conical conformal Lambert-Gaussian projection (Koldayev et al. 1978) based on phototelevision images from *Mars 4* and *Mars 5*. In drawing up the map (Fig. 6.4), techniques were developed for creating small-scale hand-drawn maps of Mars using a colour wash. The total area of the Martian surface shown amounts to about 5 million km^2. The topographic basis for the map was a mosaic of the same scale composed from phototelevision images transformed by opto-mechanical means. Above the main map are three inserts showing particular features at 1: 800,000 scale, made from the large-scale phototelevision images from *Mars 5* and transformed into the projection of the main map.

A collective of geologists from MGU, GEOKhI, VNIIzarubezhgeologiya and VNPO Aerogeologiya compiled geomorphologic and tectonic maps of Mars at 1:20,000,000 scale (Kats and Kozlov 1981). They used the 1:25,000,000 *Topographic Map of Mars* issued by the USGS in 1976 as the geographical basis.

The *Tectonic Map of Mars* consisted of nine sheets and was accompanied by a scheme of tectonic zoning for the equatorial region of Mars and a topographic map of Mars with crustal thickness contours presented on inserts at the scale of 1:50,000,000. A fragment of the tectonic map is shown in Fig. 6.5. The explanatory notes describe the tectonic features of highland, lowland and transitional regions of

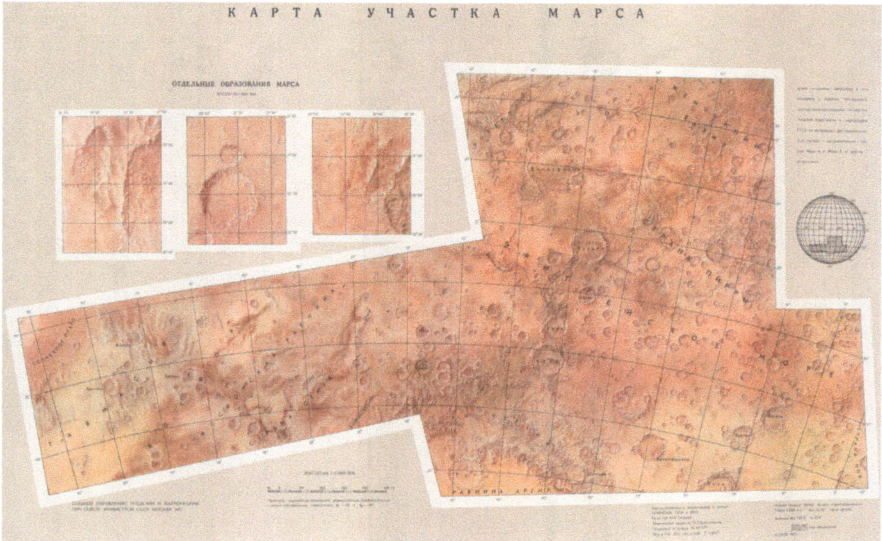

Fig. 6.4 *Map of a Region of Mars* at 1:5,000,000 scale, compiled at TsNIIGAiK using photographic material from *Mars 4, 5* in 1977

Mars. It provides information about the tectonic–magmatic activation of the planet (volcanism and rifting), ring structures, faults and other details on the tectonics. The *Geomorphological Map* at scale 1:20,000,000, also on nine sheets, identified four main terrain classes: impact and volcanic undifferentiated, volcanic, tectonogenic and exogenic. Each of these classes was divided by age and morphology and age into types, subtypes and generations. There was a regional description of the relief with detailed characteristics of individual structures of different origins and data on the relative and absolute age of the relief (Makarova et al. 1981).

In 1983, R. O. Kuzmin of GEOKhI compiled a *Map of the distribution of signs of permafrost on the surface of Mars* (Fig. 6.6). With different symbols, the map shows the polar ice caps, the ice-containing polar deposits, manifestations of thermokarst and thermoerosion, widespread manifestation of viscous plastic displacement of material on slopes, regions incised by dendritic valleys, areas of craters with fluidised ejecta, subsidence pits, major valleys and valley-like forms, landslides and other forms of relief. The insets show regions of stable persistence of ice in the surface rocks (Kuz'min 1983).

In 1989, a small production run of Mars globes of 34 cm diameter was produced, followed by a 10,000 unit production at 26 cm diameter (Shevchenko and Rodionova 1993). Using the technique of projecting cartographic images onto a spherical screen (Rodionova 1995), the USGS topographic map at 1:15,000,000 scale was transformed to the required projection for a globe. The originals for the globe segments were prepared by PKO Kartografiya. The relief illustration for the segments was made by V. D. Stushnovii. In the globe's pamphlet, there was a detailed history of the observations and mapping of Mars, information on the

Fig. 6.5 Fragment of the *Tectonic Map of Mars* (Kats and Kozlov 1981)

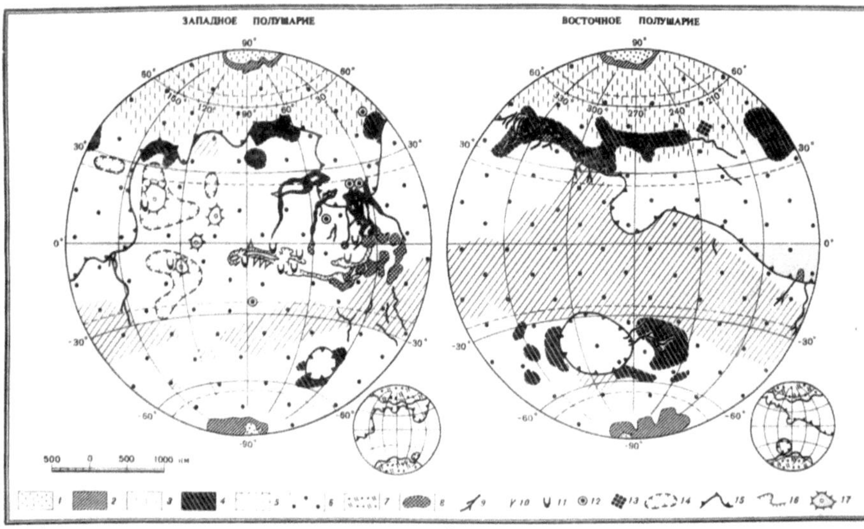

Рис. 1. Распространенность признаков проявления мерзлоты на поверхности Марса
1 – ледяной покров полярных шапок; 2 – льдосодержащие отложения полярных шапок; 3 – проявления термокарста и термоэрозии; 4 – массовое проявление вязко-пластичного смещения материала по склонам; 5 – области распространения дендритовидных долин; 6 – области распространения кратеров с флюидизированными выбросами; 7 – области устойчивого существования льда в поверхностных породах (на врезках); 8 – провально-просадочные депрессии; 9 – крупные долины; 10 – крупные оврагоподобные формы; 11 – оползни и оплывины; 12 – аласоподобные котловины; 13 – микрополигональные формы; 14 – области отсутствия кратеров с флюидизированными выбросами; 15 – границы возвышенностей и низменностей; 16, 17 – границы крупнейших каньонов и вулканов

Fig. 6.6 *Map of the distribution of signs of permafrost on the surface of Mars*, constructed by P. O. Kuz'min (1983)

planet's surface and climate and its nomenclature (Shevchenko 1989/1983). The Russian forms of the names were given according to the book by Burba (1981). Figure 6.7 shows the 26 cm globe produced by the Natural Science and School factory in 1993 and fragment of the half-tone illustration segments.

There is a special course "Mapping of extraterrestrial objects" for students of Geographical department of Moscow State University. Some students have made maps of Mars as graduation theses and are still working on this theme in Sternberg State Astronomical Institute (Rodionova et al. 2014).

An updated version of the 1989 globe was compiled in 2014 by Olga Nosova (Department of Geography Moscow University in collaboration with GAISh) with the addition of contemporary information about surface of the planet, landing sites, etc. Each of the 12 hand-made hillshade segments of the 1989 globe was converted to digital raster format, colourised and referenced to GCS_Mars_2000 in Polyconic projection. Polar segments were referenced in equidistant azimuthal projection with a 90 degree standard parallel. The Mars Global Surveyor MOLA data were smoothed, rendered and combined with the hand-shaded relief to enhance the hypsometric tints (Fig. 6.8).

The globe of Mars was produced also by a new process, shaping the 30 cm diameter hemispheres as single pieces from thermoplastic in TSNIIGAiK. It was optionally available with internal illumination.

Fig. 6.7 Mars globe at 1:26,300,000 scale, prepared by GAISh and PKO Kartografiya with a fragment of the half-tone illustration from the region of Olympus Mons

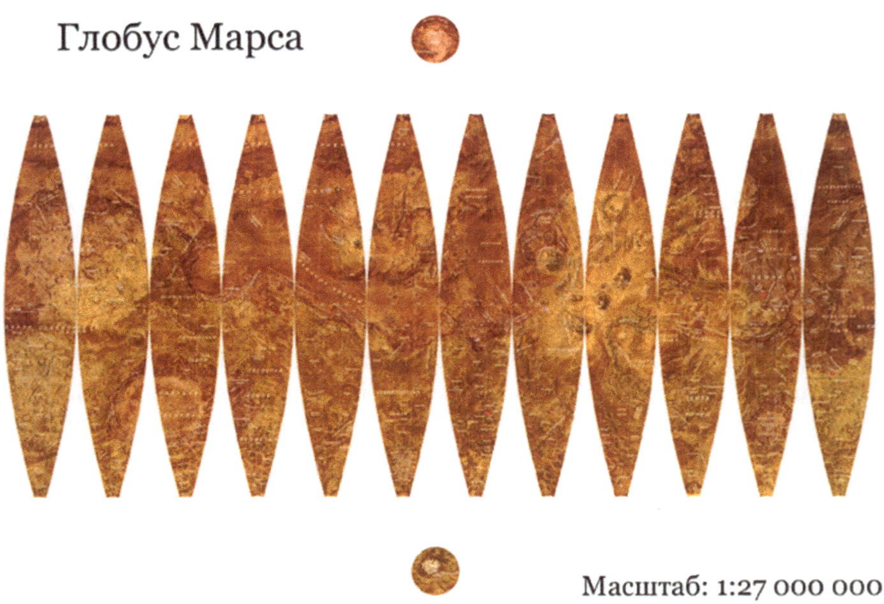

Fig. 6.8 Segments of the 2014 Mars globe

Fig. 6.9 Map of the Martian surface from *Atlas of terrestrial planets and their satellites* (1992)

In the *Atlas of terrestrial planets and their satellites* (Atlas planet zemnoy gruppy i ikh sputnikov 1992) both overview and thematic small-scale maps are presented. Figure 6.9 shows the overview *Map of the Martian surface* at 1:50,000,000 scale form the atlas, compiled in equal-area azimuthal transverse Lambert projection.

A *Hypsometric Map of Mars* at scale 1:26,000,000 (Ilukhina and Rodionova 2004) was compiled at GAISh in cooperation with the Moscow State University Department of Cartography and Geoinformatics (Fig. 6.10), using *Mars Global Surveyor MOLA* altimetry (Smith et al. 1999). Elevation is reckoned from a triaxial ellipsoid equipotential surface. The height scale contains 21 step heights: up to an elevation of +8 km the contour interval is 1 km, up to 12 km the interval is 2 km and above 12 km the interval is 10 km.

MOLA data at higher spatial resolution (1/64th degree) were used for compiling a *Hypsometric Map of Isidis Planitia* (Fig. 6.11). The mapping area was 57°–122°E

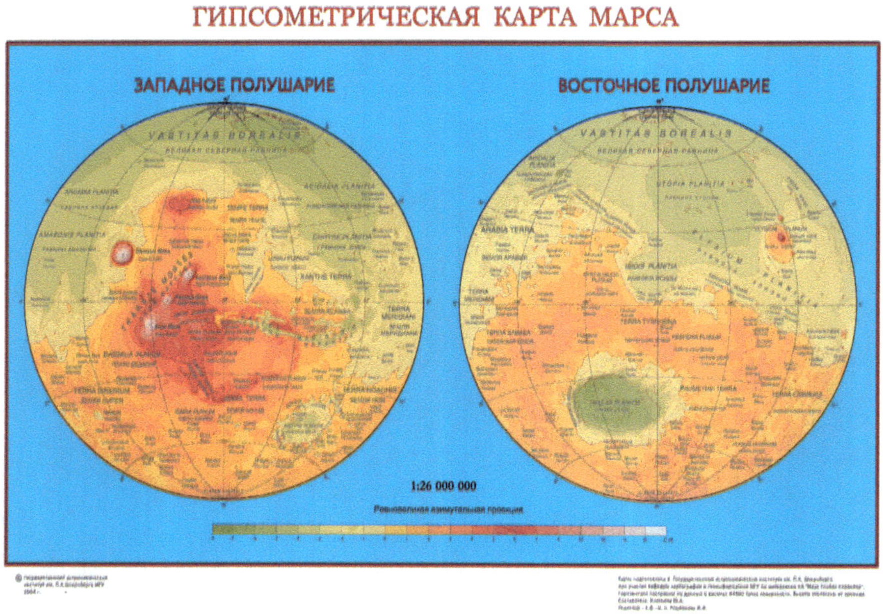

Fig. 6.10 Hypsometric map of Mars (2004)

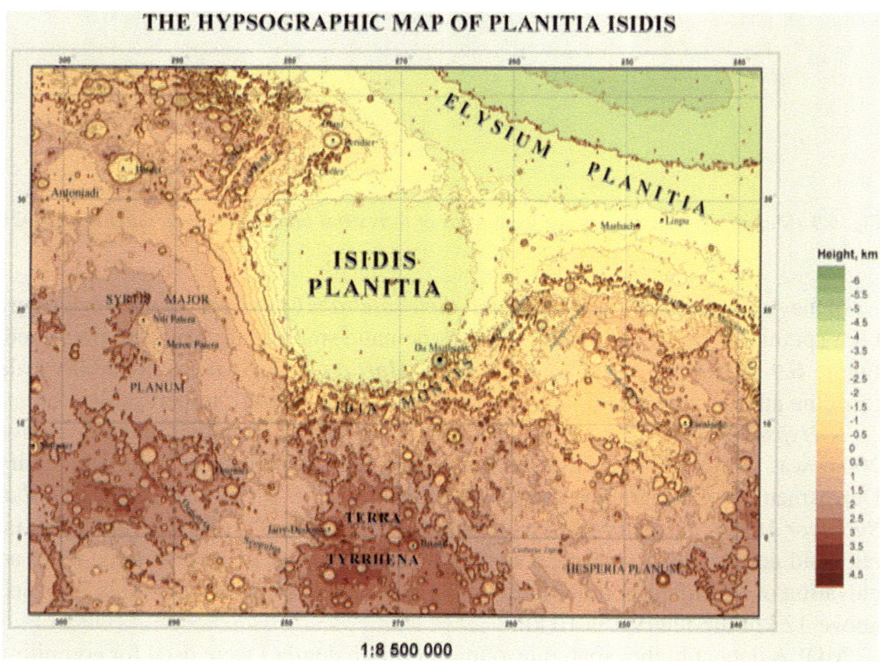

Fig. 6.11 Hypsometric Map of Isidis Planitia

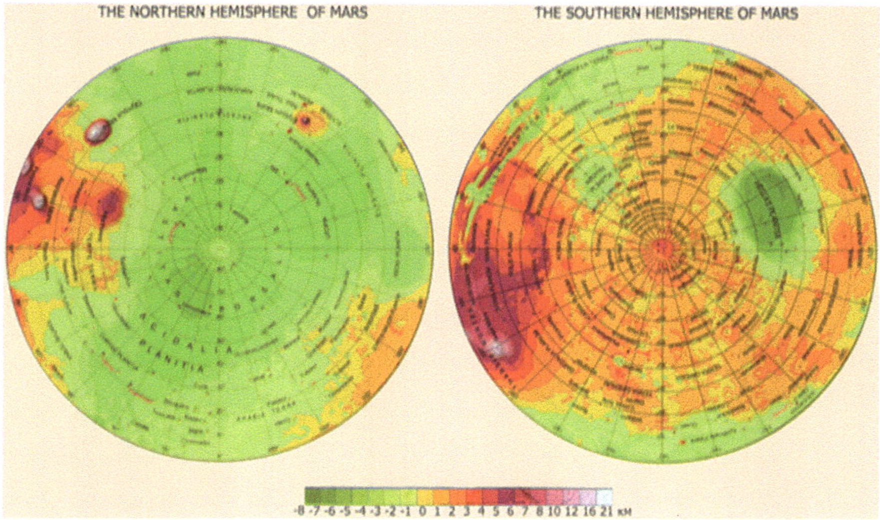

Fig. 6.12 Northern and Southern hemispheres prepared for the *Hypsometric Globe of Mars* using the thermoplastic production process

and 17°S–32°N. The DEM was processed by ArcGIS into vector contours with 500 m intervals.

The *Hypsometric Globe of Mars* of diameter 21 cm was compiled in 2012 (Rodionova and Brekhovskikh 2013) on the basis of the *Hypsometric Map of Mars* at 1:26,000,000 scale (Ilyuhina and Rodionova 2004; Ilukhina and Rodionova 2004; Rodionova and Ilukhina 2005), again using the thermoplastic production process. Western and Eastern hemispheres were recut as northern and southern hemispheres in azimuthal projection taking into account the law of deformation occurring during the deformation of the plane into a hemisphere (Figs. 6.12 and 6.13). A smaller, 15 cm version of the globe was published in 2013 (Fig. 6.14).

A variety of thematic maps have been produced reflecting the geological structure and physical characteristics of the planet's surface. Density maps for craters of different sizes and degradation states were compiled at GAISh based on the *Morphological Catalogue of the Craters of Mars* (Rodionova et al. 2000). Figure 6.15 shows, for example, the distribution of craters with ejecta larger than 10 km.

The Russian *High Energy Neutron Detector* (*HEND*) instrument launched on the NASA *Mars Odyssey* spacecraft in 2001 produced data on the surface abundance of Hydrogen: a measure of the presence of water. Figure 6.16 shows this data processed to show the relative abundance of surface ice (Mitrofanov 2005).

A *Multi-scale hypsometric map of Mars* was produced by Olga Nosova of the MGU Geography Faculty (Nosova and Lazarev 2015). The map was generated at seven different scales, each with its own hypsometric scale, and showing progressively more detail at smaller scales. In this way, the map retains a good appearance at larger scales while not compromising precision when looking at smaller areas. Such a multi-scale map can be used not only for studying the planet's relief but also as a basis for multi-scale thematic and compound maps (Fig. 6.17).

Fig. 6.13 Julia Brekhovskikh (née Ilukhina) and Zhanna Rodionova with the *Hypsometric Globe of Mars*

Fig. 6.14 *Hypsometric Globe of Mars* in 15 and 21 cm versions

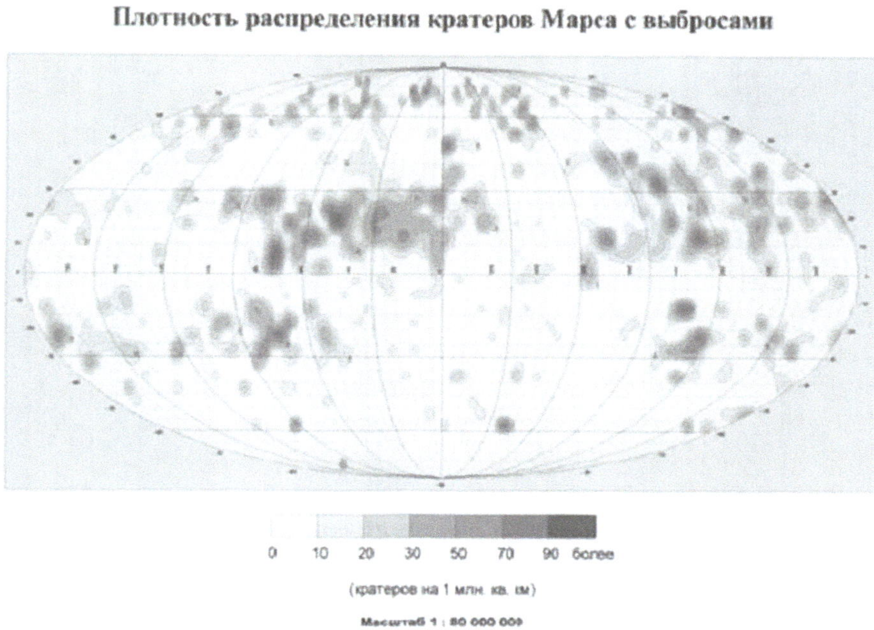

Fig. 6.15 Density distribution of martian craters larger than 10 km with ejecta

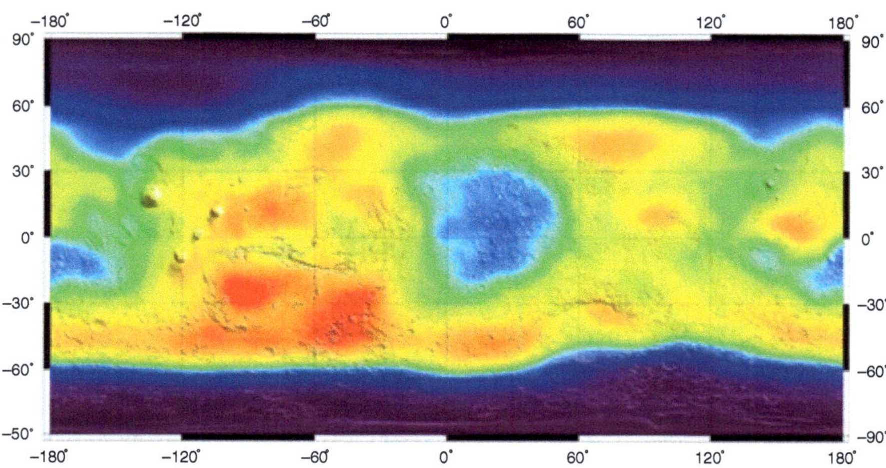

Fig. 6.16 Relative abundance of surface ice on Mars

Mapping of the Moons of Mars

Russia has a long-standing tradition in mapping and studies of the Martian satellites (Karachevtseva et al. 2014). The first Russian map of Phobos was produced jointly by MIIGAiK and Moscow State University as early as 1988 during preparation and

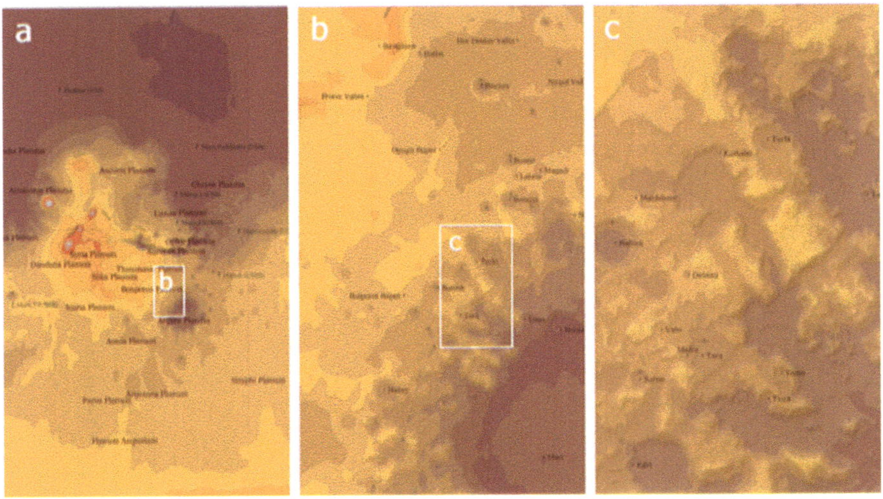

Fig. 6.17 Fragments of online *Multi-scale hypsometric map of Mars*, showing three levels of detail of relief (Nosova and Lazarev 2015)

Fig. 6.18 *Globe of Phobos* issued in 1988 by MIIGAiK and TsNIIGAiK

planning of the *Phobos 1, 2* missions. This map was produced using airbrush techniques and was based on images from the NASA *Mariner 9* and *Viking 1* missions. Also, a Phobos globe at 1:85,000 scale (Fig. 6.18) based on this map was produced to visualise the 3-dimensional morphology of the body (Bugaevsky et al. 1992).

To compile the map and globe, special map projections were developed, representing the irregular-shaped body of Phobos in the form of a triaxial ellipsoid (Bugaevsky 1987). The 1988 map was used as the basis for mapping Phobos in the *Atlas of Terrestrial planets and their satellites* (Fig. 6.19), as well as in the multilingual map series on celestial bodies (Shingareva et al. 2005).

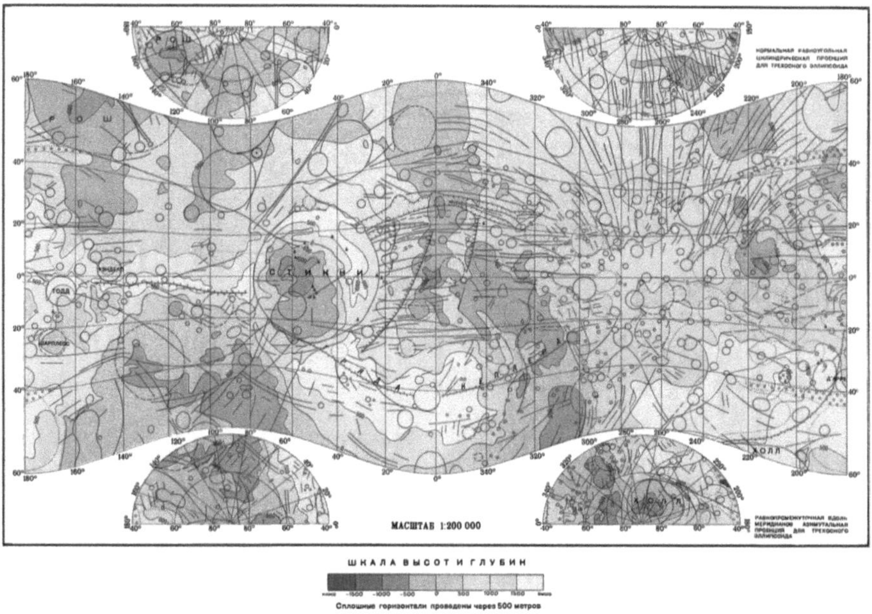

Fig. 6.19 Topographic map of Phobos from the *Atlas of Terrestrial Planets and their satellites*

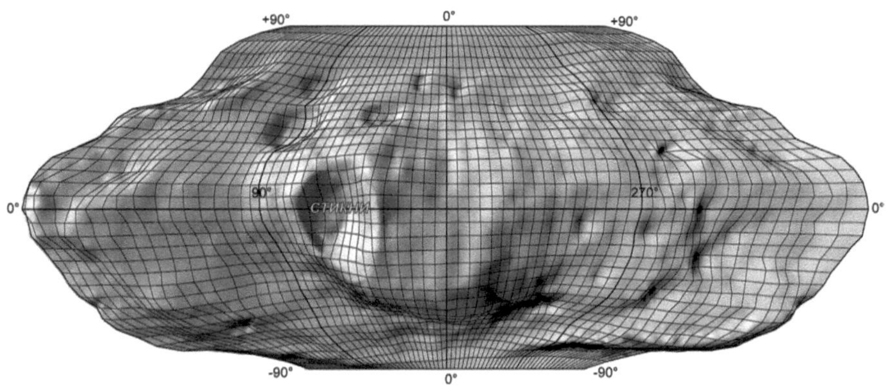

Fig. 6.20 *Map of Phobos* with calculated shaded relief (Nyrtsov 2000)

Different projections have been used to compile maps of Phobos. Figure 6.20 shows the *Map of Phobos* with calculated shaded relief (Nyrtsov 2000).

New images of Phobos were obtained as a result of the ESA Mars Express mission which obtained a large volume of image data from the *High Resolution Stereo Camera* (*HRSC*), including its *Super Resolution Channel* (*SRC*) (Oberst et al. 2008), over the course of more than 90 Phobos flybys (Witasse et al. 2014). Various maps were prepared in Russia using this data including, for example, the

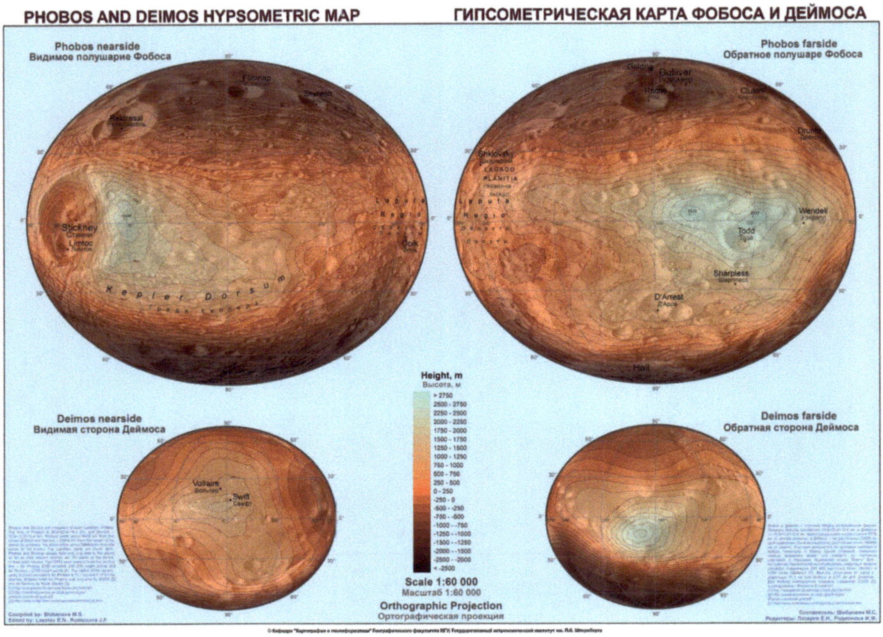

Fig. 6.21 *Hypsometric Map of Phobos and Deimos*, compiled at GAISh (Shibanova et al. 2011)

Hypsometric Map of Phobos and Deimos (Fig. 6.21) published by GAISh (Shibanova et al. 2011) which was based up on the first new shape model of Phobos derived from images by Mars Express (Willner et al. 2010).

Other examples of new Phobos mapping efforts are the recently published *Hypsometric Map of Phobos* (Karachevtseva et al. 2012) and the Phobos information system (Karachevtseva et al. 2014) which include a new control point network, a global Digital Terrain Model (DTM) as well as local terrain models and orthomosaics, derived from high-resolution images. Modern methods of image processing provide scientists with the tools for detailed spatial analyses and high quality planetary mapping. A new *Atlas of Phobos* is in preparation at MIIGAiK (Karachevtseva et al. 2015). The atlas will contain 42 Phobos thematic maps describing various aspects of the surface of the small irregular-shaped body. The maps are all based on Mars Express data except one, which is a mosaic provided by Philip Stooke (Stooke 2012b), based on higher-resolution images obtained from the *Viking* spacecraft, *Mars Global Surveyor* and *Mars Reconnaissance Orbiter*. The atlas consists of four chapters: History of Phobos mapping; Control point network, shape model and gravity field of Phobos; GIS-analyses of Phobos' surface; Geomorphologic studies of Phobos. The maps are referenced to a sphere with radius 11.1 km according to IAU recommendations (Archinal et al. 2011). Figure 6.22 shows the *Hypsometric Map* and Fig. 6.23 the *Map of Dynamic Heights* on Phobos from the atlas.

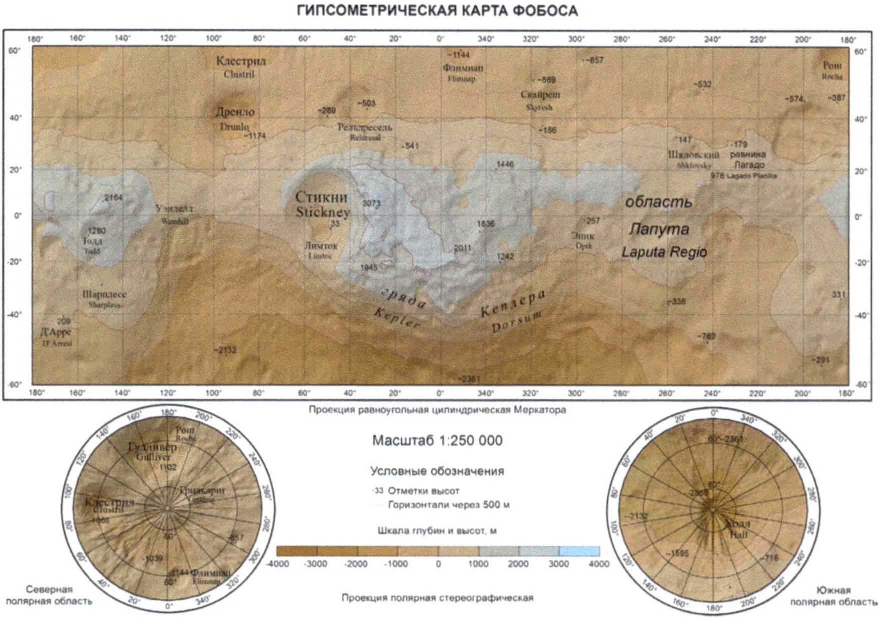

Fig. 6.22 *Hypsometric Map of Phobos* from the *Atlas of Phobos* (Karachevtseva et al. 2015)

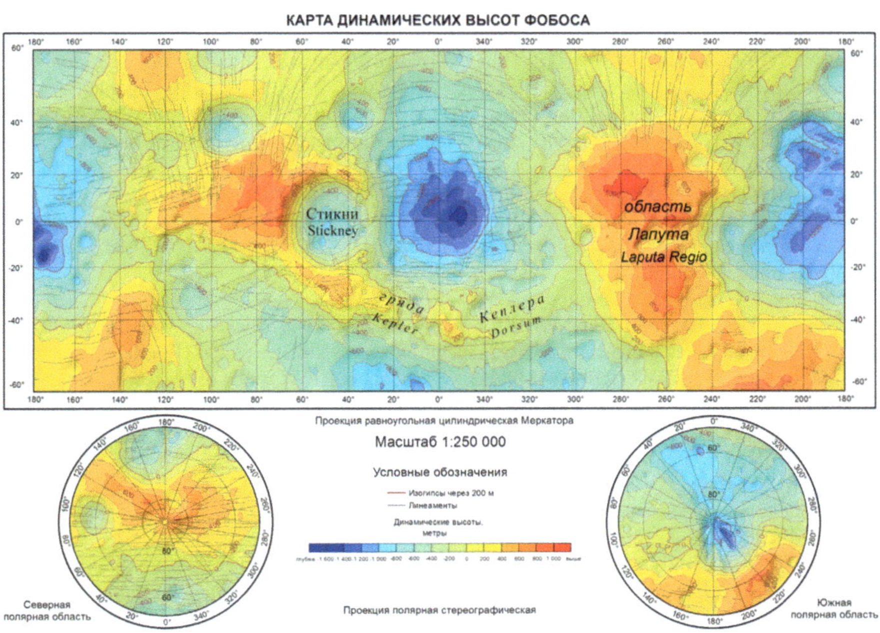

Fig. 6.23 *Map of Dynamic Heights of Phobos* from the *Atlas of Phobos* (Karachevtseva et al. 2015)

References

Atlas Planet Zemnoy Gruppu i ih Sputnikov. (1992). Izdatelstvo MIIGAiK (208 s).

Abalakin, V. K. (2005). *O.V.Struve—Dzh.V.Skiaparelli. Perepiska 1859–1904*. SPb.

Archinal, B. A., A'Hearn, M. F., Bowell, E., Conrad, A., Consolmagno, G. J., Courtin, R., et al. (2011). Report of the IAU working group on cartographic coordinates and rotational elements: 2009. *Celestial Mechanics and Dynamical Astronomy, 109*.

Blunck, J., Zögner, J., & Jöns, H.-P. (1993). *Der Rote Planet im Karten Bild: 200 Jahre Marskartographie von Herschel, Beer und Mädler bis zur CD*. Gotha.

Bugaevsky, L. M. (1987). On the problem of elaborating of isometric coordinates and equiangular cylinclric projection of the tri-axial ellipsoid. IzvestiyaVUZov, ser. *Geodesy and Aerial Photography, 4*, 79–90.

Bugaevsky, L. M., Krasnopevtseva, B. V., & Shingareva, K. B. (1992). Phobosmapand Phobos globe. *Advances in Space Research, 12*(9), 17–21. 10.1016/0273-1177(92)90314-N.

Burba, G. N. (1981). *Nomenklatura detaley rel'yefa Marsa* (87 s). Moskva: Izdatelstvo Nauka.

Ilukhina, Yu. A., & Rodionova, J. F. (2004). Automated making of the map of Isidis Basin. *Proceedings of the Conference Devoted 225 years of MIIGAiK* (pp. 402–407).

Ilukhina, Yu. A., & Rodionova, Zh. F. (2004). *Gipsometricheskaya karta Marsa. 1:26 000 000. M.*

Karachevtseva, I. P., Shingareva, K. B., Konopikhin, A. A., Mukabenova, B. V., Nadezhdina, I. E., & Zubarev, A. E. (2012). GIS mapping of phobos on the results of data processing of remote sensing satellite mars express. Modern problems of remote sensing of the earth from space (pp. 304–311). Moscow: Space Research Institute (in Russian).

Karachevtseva, A., Kokhanov, J., Rodionova, A., Konopikhin, A., Zubarev, I., Nadezhdina, L., et al. (2015). Development of a new Phobos atlas based on Mars express image data. *Planetary and Space Science, 108*, 24–30.

Karachevtseva, I. P., Oberst, J., Zubarev, A. E., Nadezhdina, I. E., Kokhanov, A. A., Garov, A. S., et al. (2014). The Phobos information system. *Planetary and Space Science, 102*, 74–85. 10.1016/j.pss.2013.12.015 (Special IssuePhobos).

Kats, Ya. G., & Kozlov, V. V. (1981). *Sulidi-Kondrat'yev i dr. Ob"yasnitel'naya zapiska k tektonicheskoy karte Marsa masshtaba 1:20 000 000 M* (39 s). VNIIzarubezhgeologiya.

Koldayev, P. K., Ledovskaya, L. S., Ostrovskiy, M. V., Tyuflin, Yu. S., & Fokina, L. A. (1978). *Karta uchastka Marsa masshtaba 1:5 000 000*. Sb. Kartografirovaniye Luny i Marsa. M. (s.75-78).

Kuz'min, R. O. (1983). *Kriolitosfera Marsa* (141 s). Moskva: Izdatelstvo Nauka.

Makarova, N. V., Kats, Ya. G., & Kozlov, V.V. (1981). *i dr. Ob"yasnitel'naya zapiska k Geomorfologicheskoy karte Marsa masshtaba 1:20 000 000. M* (38 s). VNIIzarubezhgeologiya.

Mitrofanov, I. G. (2005). Poiski vody na Marse. *Priroda, 9*, S. 24–42.

Nosova, O. S., & Lazarev, E. N. (2015). *Multi-scale hypsometric map of Mars*. https://a.tiles.mapbox.com/v4/lekanosova.mbh8l3ce/page.html?access_token=pk.eyJ1IjoibGVrYW5vc292YSIsImEiOiJQSUZWR1RJIn0.s4M9uC3hOLRn5qqeu0RPWQ#1/5/-89

Nyrtsov, M. V. (2000). *Developing map projections of actual surfaces of celestial bodies and methods of their study* (Ph.D. thesis). Moscow State University of Geodesy and Cartography (MIIGAiK).

Oberst, J., Schwarz, G., Behnke, T., Hoffmann, H., Matz, K.-D., Flohrer, J., et al. (2008). The imaging performance of the SRC on Mars Express. *Planetary and Space Science, 56*, 473–491.

Rodionova, Zh. F., Berlyant, A. M., Brekhovskikh, J. A., Lazarev, E. N., Lazareva M.S., Nosova, O. S. (2014). *Maps of Mars compiled by students in Lomonosov Moscow State University* (pp. 319–326). Thematic Cartography for the Society, Switzerland.

Rodionova, Zh. F., & Brekhovskikh, J. A. (2013). Hypsometric globe of Mars—3D model of the planet. *Proceedings ICC2013* (s. 599–603). Dresden.

Rodionova, Zh. F., Chikmachev, V. I., & Shevchenko, V. V. (1995). Obzornaya kartograficheskaya model' poverkhnosti Marsa. *Astronomicheskiy vestnik, tom 29*(5), s. 405–412.

Rodionova, J. F., Dekhtyareva, K. I., Khramchikhin, A. A., Michael, G. G., Ajukov, S. V., Pugacheva, S. G., et al. (2000). In V. V. Shevchenko & A. F. Chicarro (Eds.), *Morphological catalogue of the craters of Mars* (158 p). ESA/ESTEC Noordwijk.

Rodionova, Zh. F., & Ilukhina, Yu. A. (2005). Novaya karta Marsa. *Zemlya i Vselennaya, 2*, S. 38–48.

Sagan, C., & Fox, P. (1975). The canals of Mars: An assessment after Mariner 9. *Icarus, 25*, 602–612.

Shevchenko, V. V. (1989/1993). Globus Marsa 1:19 500, 1:26 300 000. GAISH MGU, PKO «Kartografiya» GUGK./Sost. V.I.Stushnova, red. ZH.F.Rodionova, nauchn. ruk.

Shevchenko, V. V., & Rodionova, Zh. F. (1993). *Globus Marsa—yeshche odna «planeta» u vas na stole* (s. 101–135). Moskva.

Shibanova, M. S., Lazarev, E. N., & Rodionova, J. F. (2011). *Phobos and Deimos hypsometric map M.* Moskva: Pechatniki.

Shingareva, K. B., Krasnopevtseva, B. V., Leonenko, S. M., Fleis, M. E., Buchroithner, M. F., Waelder, O., et al. (2005). Phobos and Deimos—A new map in the series of multilingual relief maps of terrestrial planets and their moons. In: *Proceedings of 22nd International Cartographic Conference*, LaCoruna, Spain (CD-ROM).

Smith, D. E., Zuber, M. T., Solomon, S. C., Phillips, R. J., Head, J. W., Garvin, J. B., et al. (1999). The global topography of Mars and implications for surface evolution. *Science, 284*, 1495–1503.

Stooke, P. J. (2012a). *The international atlas of mars exploration. The first five decades: 1953–2003*. Cambridge: Cambridge University Press.

Stooke P. (2012b). Stooke small bodies MapsV2.0.MULTI-SA-MULTI-6-STOOKEMAPS-V2.0. NASAPlanetary Data System.

Willner, K., Oberst, J., Hussmann, H., Giese, B., Hoffmann, H., Matz, K.-D., et al. (2010). Phobos control point network, rotation, and shape. *Earth and Planetary Science Letters, 294*, 547–553.

Witasse, O., Duxbury, T., & Chicarro, A. (2014). Mars express investigations of Phobos and Deimos. *Planetary and Space Science, 102*, 18–34. Accessed 14 Jan, 2015, from 10.1016/j.pss.2013.08.002 (Special Issue Phobos).

Chapter 7
Cartography of Venus

Abstract It is not possible to see the surface of Venus even from orbit around the planet because it is covered by a thick, cloudy atmosphere. Venus has been studied by over 20 spacecraft including the *Venera, Mariner, Vega* and *Pioneer Venus* programmes and *Magellan* and *Venus Express*. The Soviet contribution to Venus research was particularly significant, making the first soft landing and acquiring the first panoramas from its surface. *Venera 15, 16* were equipped with a side-looking radar which mapped more than 30 % of the northern hemisphere at a higher resolution than had *Pioneer Venus*. On the basis of these materials, an *Atlas of the Surface of Venus* was compiled in the USSR including photomaps, hypsometric, geomorphological and other special maps. Later hypsometric maps were prepared at GAISh in 2008, 2010 and 2012 on the basis of the *Magellan* altimetry data.

Venus is among the brightest objects of the sky. Its brightness is partly explained by the reflection of the solar light from its thick, cloudy atmosphere. Because of this atmosphere, it is not possible to see the surface even from orbit about the planet. As other planets, Venus orbits the Sun in an anticlockwise direction, taking 225 days to do so: its rotation period about its own axis of 243 days was determined only in the 1960s with the early use of radar techniques. From the Earth-based radar observations, two bright relief features were identified and named, from the Greek alphabet, Alpha and Beta, and from these it was possible to discover the rotation period of the planet. Unlike other planets, where the orbital and rotational directions are the same, Venus rotates oppositely from its orbit—clockwise, and for this reason, longitudes there are counted off from west to east.

Venus's atmosphere was discovered by the Russian scientist, Mikhail Vasilyevich Lomonosov (1711–1765), in 1761 at the time of a Venus transit in front of the disc of the Sun. As Venus moved off the solar disc, he noticed a tiny bulge at the edge of the Sun. He correctly proposed that this effect was caused by the bending of the solar rays through the atmosphere of the planet (Fig. 7.1).

Observations with the 12.6 cm radar system at the Arecibo Observatory in 1975 and 1977 yielded images covering approximately 25 % of the surface of Venus. The resolutions of these images ranged from 5 to 20 km (Campbell and Burns 1980). Radar was used both from the Earth and from spacecraft, the proximity of spacecraft offering much higher resolution of surface details. The first such experiment was carried on the NASA spacecraft *Pioneer Venus* in 1978–1979 and later on the

V. Shevchenko et al., *Lunar and Planetary Cartography in Russia*, Astrophysics and Space Science Library 425, DOI 10.1007/978-3-319-21039-1_7

Fig. 7.1 Mikhail Lomonosov and his sketches of the transit of Venus, explaining his conclusions about the presence of an atmosphere around the planet

Soviet *Venera 15, 16* (1983) and NASA *Magellan* (1990–1994) (Ford 1992; Hare 2002).

Soviet scientists made a particularly significant contribution to the study of Venus. The Soviet spacecraft *Venera 1* made a flyby at a distance of 100,000 km from the planet in spring 1961, followed by the American *Mariner 2* in December 1962 at 35,000 km, which determined the temperature of the surface. It also discovered that the planet has no magnetic field or radiation belts. *Venera* 2 was also a flyby spacecraft, passing at a distance of 24,000 km in February 1966.

In October 1967, the *Venera 4* spacecraft conducted scientific measurements during its descent through the atmosphere and was the first to establish the atmospheric composition. This was the first success in the direct study of Venus by spacecraft. However, the maximum pressure for which the sensors had been designed on *Venera 4* was 7.2 bar, which was not enough: at an altitude of 23 km, when the pressure reached 17.6 bar, the apparatus collapsed. However, comparison of the *Venera 4* data on altitude and pressure with that from the American *Mariner 5*, which made a flyby a day later, made it possible to calculate the pressure at the planet's surface: about 100 bar. *Venera 5* and *6* were built to be more robust and operated down to an altitude of about 17 km, where the pressure reached 27 bar.

The first apparatus to reach the surface was *Venera 7* in 1970. In 1972, *Venera 8*, designed for a pressure of 120 bar, made a soft landing. These missions included an service module, which carried scientific and service instruments, as well as the descent module. Approaching the planet, the spacecraft would receive the command to separate the descent capsule from the service module, which is then destroyed in the upper atmosphere, and the lander, equipped with a special coating, would decelerate in the atmosphere and then descend on a parachute. *Venera 8* registered data on the surface temperature and pressure, which proved to be very high: 472 °C and 93 bar.

Fig. 7.2 Colour image of the surface of Venus at the landing site of *Venera 13*, which operated to the east of Phoebe (7.5°S 303°E) for 2 hours and 7 minutes on 1st March 1982

In February 1974, the spacecraft *Mariner 10* flew past Venus imaging its cloud cover over a period of 8 days to study the dynamics of the atmosphere.

An important milestone in the study of the planet and great successes of the Soviet space program occurred on 22 and 25 October 1975, when *Venera 9* and *10* went into orbit as the first artificial satellites of the planet, and their landers transmitted the first images of its surface. Both spacecraft landed in Beta Regio: *Venera 9* in the northern part and *Venera 10* in the south. They also carried a set of scientific instruments which included a scanning UV telephotometer, which transmitted images of the cloud layer above. The wind speed at the surface was measured to be 0.4–1.3 m/s.

Further research broadened our knowledge of Venus step by step. In 1978, *Venera 11* and *12* conducted a study of solar radiation and the composition of the Venusian atmosphere, among other things. In 1982, before the construction of *Venera 13* and *14,* a very complex goal was set: to obtain the first colour images of the planet from their landing sites (Fig. 7.2). These craft were also the first to carry out a chemical analysis of the soil. They were targeted much further south than the Beta Regio: 750 and 1350 km south of the equator, to the east of the Phoebe mountain range.

Three years later, in 1985, the spacecraft *Vega 1* and *2* brought two relatively small 3.4 m diameter balloons to the atmosphere of Venus, to study the planet's wind. The *Vega 1* balloon began its journey 8° north of the equator and *Vega 2*'s balloon 7.5° to the south. For two days, the balloons floated freely in the atmosphere at a height of 53–54 km, sending signals back to the Earth. The landers, meanwhile, conducted a study of the atmosphere and the planet's surface. The region of the landing sites—which were close together—later became known as the Rusalka Planitia (a rusalka is a mermaid in Russian). The landing sites of the *Venera* and *Vega* landers are shown on the map compiled by the US Geological Survey from the *Pioneer Venus* data (Fig. 7.3).

The Soviet planetary orbiters *Venera 15* (Fig. 7.4) and *16* (1983–1984) were equipped with side-looking radar, enabling them to map more than 30 % of the surface of the northern hemisphere with a resolution exceeding that of the elevation data obtained by *Pioneer Venus*. As a consequence of the orbital motion of the spacecraft, the radar recorded data from a swath parallel to the line of flight.

The spacecraft were imaging from highly elongated polar orbits, moving from the pole to the equator at closest approach. Consequently, each close approach provided an image area spanning a width of about 160 km and a length of 8000 km.

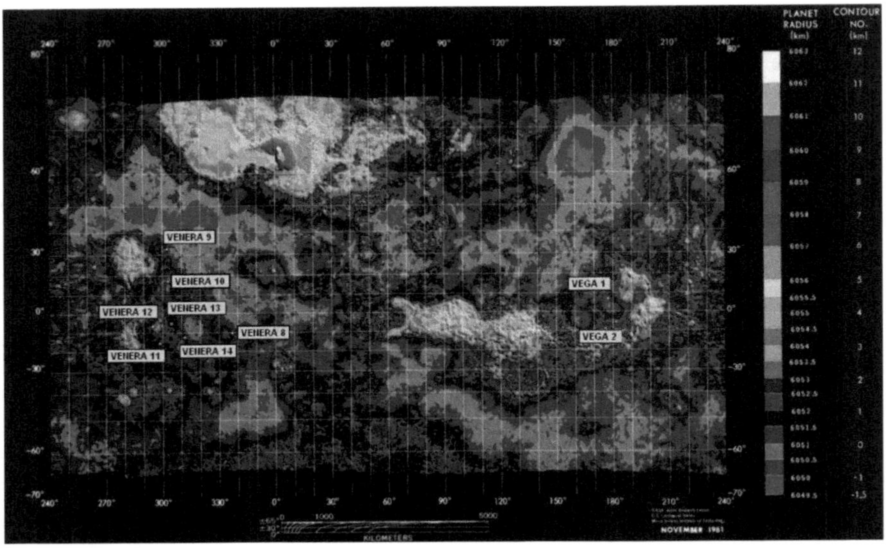

Fig. 7.3 Landing sites of the *Venera* and *Vega* landers on the map compiled by the US Geological Survey from the *Pioneer Venus* data

Between orbits, the planet turned enough so that the next imaging strip just overlapped with the previous one, eventually providing solid coverage over an area of 115 million km^2 with a resolution of 1–2 km. An *Atlas of the surface of Venus* was compiled from these materials, including photomaps, hypsometric, geomorphological and other special maps at 1:10,000,000 scale, as well as photoplans at 1:4,000,000 scale. The *Atlas of the surface of Venus* was published in 1989 in the USSR with a print run of 1110 copies (Atlas Poverhnosti Venery 1989) (Fig. 7.5). 27 sheets of geomorphological maps compiled at GEOKhI were prepared for publication by PKO Kartografiya and GUGK Enterprise No 7. A description of the parts of the surface of Venus covered by the atlas was published by Basilevskij and Janle (1987).

From the data of *Pioneer Venus* and *Venera 15, 16,* a hypsometric map of Venus was compiled for the *Atlas of the Terrestrial Planets and their Satellites* at 1:75,000,000 scale in transverse azimuthal equal area Lambert projection. Contours were drawn at 1 km intervals (Atlas Planet Zemnoy Gruppu 1992). The atlas contained several additional thematic maps of Venus (Fig. 7.6).

The data from *Venera 15, 16* were also used for the production of a series of maps by the USGS in collaboration with GEOKhI. The maps were constructed in polar stereographic projection for the northern hemisphere of Venus down to 20°N with a scale ranging from 1:15,000,000 at 40°N to 1:18,250,000 at the north pole. The series included three types of map: a radar image map of the relief, a topographic map with contours and layered colours and a geological map showing 12 types of terrain differing in origin and age. The most widespread lava plains with low sinuous ridges were depicted in green, the oldest highlands–tessera in dark pink

Fig. 7.4 *Venera 15* spacecraft

and the youngest lava flows in yellow. Dark blue showed the belt of mountain ridges, which have an intermediate age. The north pole was at the centre of the map. The maps covered 21 % of the surface of Venus, corresponding to a territory of 98 million km^2, equivalent to the combined area of Asia, Africa and North America.

Hypsometric (topographic) maps play an important role in the thematic mapping of the planets and their moons, since the formation processes of the relief of hard-surface celestial bodies is a primary focus of research. Relief maps provide a foundation for many thematic maps of interest in planetary astronomy and planetary geodesy (Rodionova 1984; Rodionova and Dekhtyareva 1986).

The data received from *Pioneer Venus* (Pettengill and Eliason 1980), *Venera 15, 16* (Rzhiga 1987) and *Magellan* (Batson et al. 1994; Wu and Howington-Kraus 1994) permitted the production of hypsometric maps with increasing accuracy. In 2008, a *Hypsometric Map of Venus* (Fig. 7.7) showing the hemispheres at 1:90 M

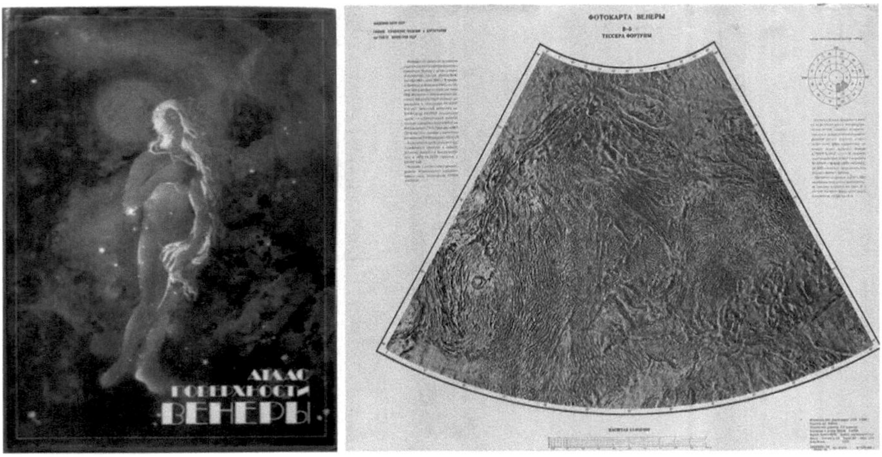

Fig. 7.5 Cover of the *Atlas of the surface of Venus* and a sheet of the photomap covering Lakshmi Planum

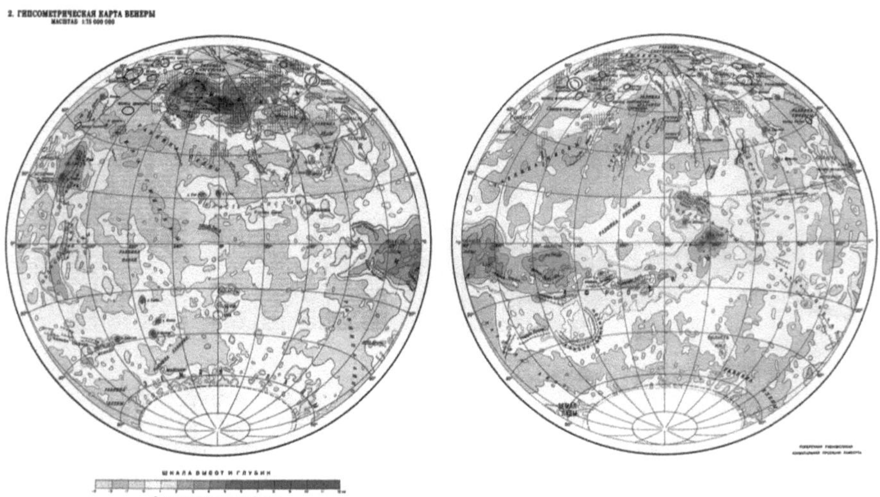

Fig. 7.6 Hypsometric map of Venus from the *Atlas of the Terrestrial Planets and their Satellites*

scale was compiled at GAISh (Lazarev and Rodionova 2008). The map was prepared on the basis of *Magellan* altimetry data (Batson et al. 1994; Wu and Howington-Kraus 1994), referenced to a sphere of radius 6051.0 km (Lazarev and Rodionova 2006). The central meridian of the left hemisphere is at 0°E and of the right at 180°E.

The map was constructed in Lambert azimuthal equal-area projection for easy visualisation of the hemispheres of the planet. Contours were calculated by averaging the elevation data over one-degree trapezoids. In areas of maximum and

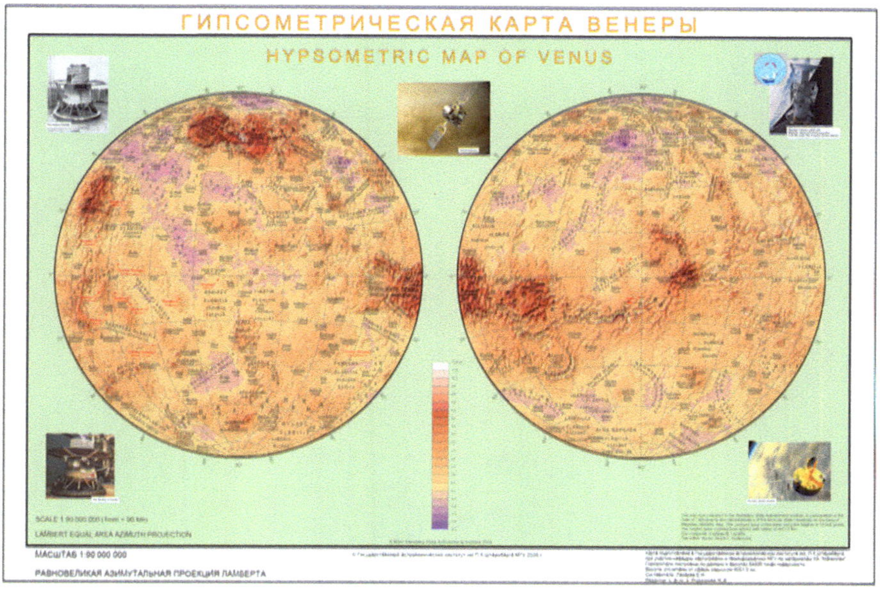

Fig. 7.7 Hypsometric map of Venus (Lazarev and Rodionova 2008)

minimum heights, such as the Maxwell Montes, Maat Mons and Atalanta Planitia, the trapezoid size was reduced to improve the accuracy of the depicted areas. Latin names were applied to the map according to the IAU nomenclature.

A new relief map at 1:45 M scale was compiled and published in 2010 (Fig. 7.8). The compilation techniques were refined from those of the 2008 map. A sphere of radius 6051.8 km was used as the reference surface (Seidelmann et al. 2007). The *Magellan* radar (SAR) spot was 5×5 km: for the map it was sufficient to use a DTM with 0.07 degrees between height samples, equivalent to 7.5 km at equator. So we averaged the source DTM with minimal sacrifice of accuracy onto a new grid of 7.5×7.5 km. Polygons were created using the spline interpolation method. Contours and hillshade were added, and finally, feature names and other information.

The location of the central meridians of the hemispheres were selected on the basis that, every 583 days, Venus is positioned in conjunction with the Earth. In this position, the 320°E meridian is directed towards the Earth, and 140°E is directed away. This situation will hold for at least 600 years (Burba 1996). Thus, 320° and 140°E were taken as central meridians for the hemispheres, corresponding to the near and far sides at closest approach to Earth. At the same time, this choice is convenient for representation of Aphrodite Terra, the largest terra on Venus, because it then falls into a single hemisphere. Two additional inset views of the polar regions were included at the same scale. For this map as a whole, a special colour scale was used to represent the surface elevation, where lowlands below the zero level have a violet colour, and higher terrain is shown with yellows, browns

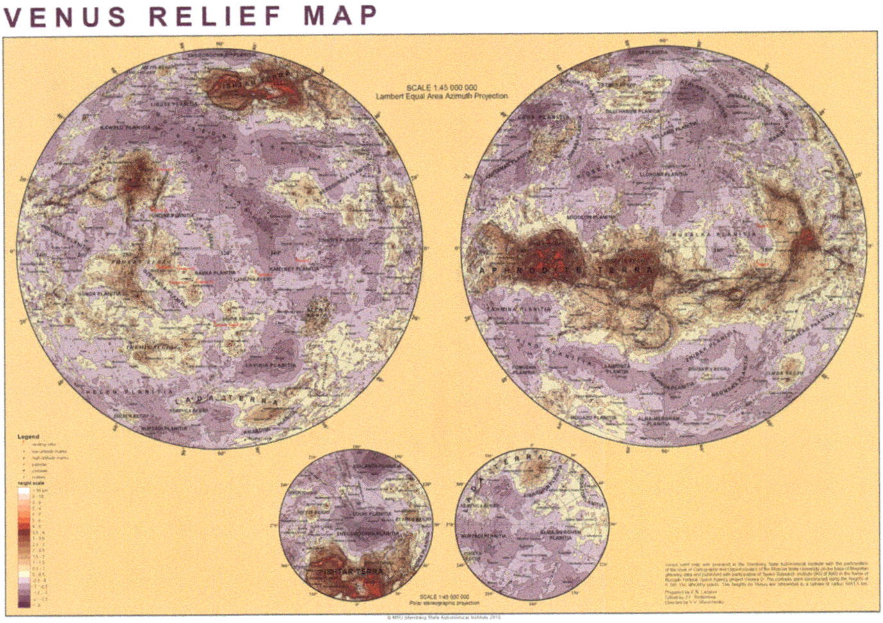

Fig. 7.8 Venus relief map (Lazarev et al. 2010)

and reds. The coordinate grid was drawn at 20° intervals. The precision of representation of the relief of the new map was higher than for the map published in 2008. The *Venus relief map* with Latin names was published with the financial support of IKI. Based on this map, a *Hypsometric Globe of Venus* (Fig. 7.9) was also produced (Brekhovskikh et al. 2012).

A new version of the *Venus relief map* was produced in 2012 including both Russian and Latin names, in two colour variants (Fig. 7.10).

The nomenclature of features on Venus, as on other planets, is defined by the IAU (http://planetarynames.wr.usgs.gov) only in Latin script, so that planetary cartography in Russia requires an appropriate Russian transcription of the names. On the other hand, the processing of materials from planetary spacecraft in the Soviet space programme established many names first in Russian, requiring elaboration of their Latin equivalents for international adoption (Burba 1988).

The working group of the IAU on the nomenclature of Venus, consisting of representatives of the USSR and the United States, agreed to give surface features of the planet only female names: mountains are given the names of goddesses, and lowland relief features the names of other mythological characters. However, the names of the three bright features seen in early radar images—Alpha, Beta and Maxwell—were retained and got new life as the names of highland features—Alpha Regio, Beta Regio and Maxwell Montes, the only exception in favour of a man.

Fig. 7.9 *Hypsometric Globe of Venus* (Brekhovskikh et al. 2012) produced from the *Venus Relief Map* (Photo: V. Surdin)

Feature names in Russian reverse the order of the words: in contrast to the Latin version, where the first word is a generic term and the second a proper name, for example, Beta Regio becomes oblast' Beta, or Snegurochka Planitia becomes ravnina Snegurochki. This corresponds to the conventional construction of Russian geographical names. Another feature is that, unlike the Latin terms, which use capital letters for both the name and the generic term, the generic terms in Russian are usually written with a lowercase letter (Burba 1988). These differences are important in the preparation of maps which show the names in both languages.

On the 2012 map, landforms occupying vast areas (terrae, planitiae, plana) are marked with upper case letters and linear landforms (dorsa, chasmata) in lower case. In order to show the difference and diversity of the types of landform, many of them have their own font styles (size, slope). Since the Latin names are the approved IAU designations, they are printed larger than the Russian transcriptions

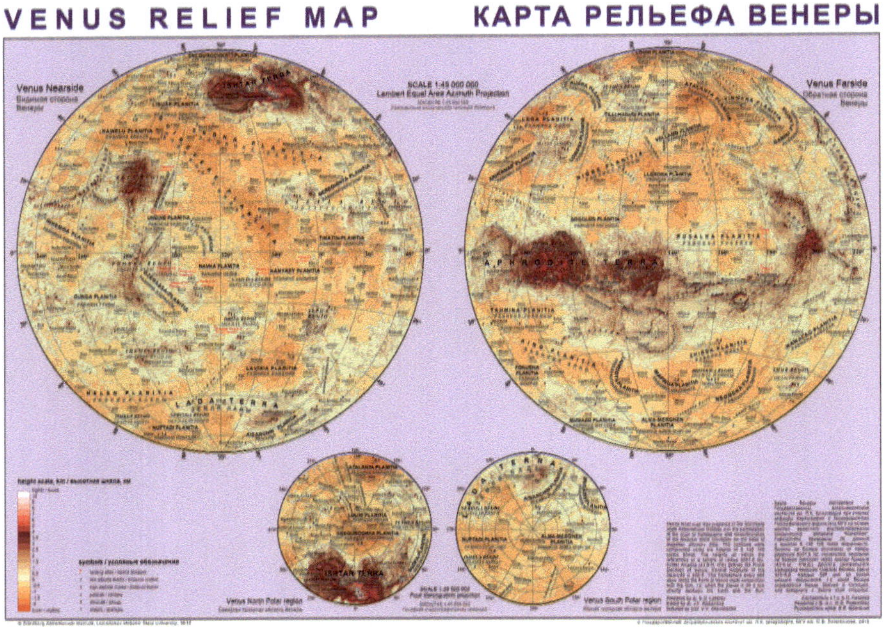

Fig. 7.10 Venus relief map with Latinian and Russian names (Lazarev et al. 2012)

(Lazarev and Rodionova 2012). Latin names of craters larger than 100 km are marked with a larger font. Craters are named after prominent women such as Cleopatra (65.8°N 7.1°E, 105 km in diameter) and Akhmatova (61.3°S 307.9°E, 41.4 km). Small craters take womens' first names, such as Anya or Irina. In addition to named surface features, spacecraft landing sites were also marked and the surface elevation given for various points of interest.

References

Atlas Planet Zemnoy Gruppy i ih Sputnikov. (1992). MIIGAiK, Moskva.

Atlas Poverhnosti Venery. (1989). In V. A. Kotelnikov, V. R. Yashchenko, & A. F. Zolotov (Eds.). GUGK, Moskva.

Basilevskij, A. T., & Janle, P. (1987). Geologic-morphologic and gravimetric characteristics of Bell Regio on Venus. *Astronomicheskij vestnik, XXI*(2), 109–121.

Batson, R. M., Kirk, R. L., Edwards, K. F., & Morgan, H. F. (1994). Venus cartography. *JGR, 99*, 21173–21182.

Brekhovskikh, J., Lazarev, E., & Rodionova, J. (2012). *Hypsometriy globus o Veneru*. Moscow: IKI, GAISh MSU.

Burba, G. A. (1988). *Nomenklatura detaley rel'yefa Venery. M.* Nauka (64 c).

Burba, G. A. (1996). *Cartographic aspects of Venus global geologic mapping at 1:10,000,000 scale*. Vernadskiy-Brown Micro 24 abstracts 11.

Campbell, D. B., & Burns, B. A. (1980). Earth-based radar imagery of Venus. *Journal of Geophysical Research, 85*, 8271–8281.

Ford, P. G. (1992). *ARCDRLST—A program to list Magellan altimetry and radiometry composite data records*. Center for Space Research Massachusetts Institute of Technology, Version 2.2.

Hare, T. M. (2002). *GTDRtoporraphy: Image and contour information*. Flagstaff, AZ: U.S. Geological Survey Astrogeology.

Lazarev, E. N., & Rodionova, J. F. (2006). *Automatic creation of the hypsometric map of Venus*. Brown-Vernadsky Micro 44 abstracts, 53–54.

Lazarev, E. N., & Rodionova, J. F. (2008). *Hypsometric map of Venus, scale 1:90 000 000*. Moscow.

Lazarev, E. N., Rodionova, Zh. F., & Shevchenko, V. V. (2010). *Venus relief map, scale 1:45 000 000*. Moscow.

Lazarev, Ye. N., & Rodionova, Zh. F. (2012). *Kartografirovaniye Venery* (116 s). Saarbrucken: LAP Lambert Academic Publishing. ISBN 978-3-659-25082-8.

Lazarev, E. N., Rodionova, Zh. F., & Shevchenko, V. V. (2012). *Venus Relief Map 1:45 000 000*. Moscow: Sternberg Astronomical Institute, Moscow University. Izdatelstvo Print.

Pettengill, G. H., & Eliason, E. (1980). Pioneer Venus radar results—Altimetry and surface properties. *JGR, 85*, 8261–8270.

Rodionova, Zh. F. (1984). Gipsometricheskiye osobennosti Venery. *Astronomicheskiy vestnik, 18* (2), S. 117–127.

Rodionova, Zh. F., & Dekhtyareva, K. I. (1986). *Gipsometricheskiye osobennosti Luny i planet zemnoy gruppy*. Problemy kompleksnogo issledovaniya Luny; M.: Izd. MGU (S. 56–71).

Rzhiga, O. N. (1987). *Venera-15 and -16 spacecraft—Images and maps of Venus*. COSPAR, IAU, Plenary Meeting, 26th vol. 7 12: 269–278.

Seidelmann, P. K., Archinal, B. A., A'Hearn, M. F., Conrad, A., Consolmagno, G. J., Hestroffer, D., et al. (2007). Report of the IAU/IAG working group on cartographic coordinates and rotational elements 2006. *Celestial Mechanics and Dynamical Astronomy, 98*, 155–180.

Wu, S. S. C., & Howington-Kraus, E. A. (1994). *Magellan radar data for Venus topographic mapping*. Abstracts of the 25th Lunar and Planetary Science Conference 1519.

MIX
Papier aus verantwortungsvollen Quellen
Paper from responsible sources
FSC® C105338

If you have any concerns about our products,
you can contact us on
ProductSafety@springernature.com

In case Publisher is established outside the EU,
the EU authorized representative is:
Springer Nature Customer Service Center GmbH
Europaplatz 3, 69115 Heidelberg, Germany

Printed by Libri Plureos GmbH
in Hamburg, Germany